Charles Creighton

The Natural History of Cow-Pox and Vaccinal Syphilis

Charles Creighton

The Natural History of Cow-Pox and Vaccinal Syphilis

ISBN/EAN: 9783337184148

Printed in Europe, USA, Canada, Australia, Japan

Cover: Foto ©berggeist007 / pixelio.de

More available books at **www.hansebooks.com**

THE NATURAL HISTORY OF COW-POX

AND

VACCINAL SYPHILIS.

THE

Natural History of Cow-Pox

AND

Vaccinal Syphilis.

BY

CHARLES CREIGHTON, M.D.

. . . . " Incedis per ignes
Suppositos cineri doloso."—*Horace.*

CASSELL & COMPANY, LIMITED:

LONDON, PARIS, NEW YORK & MELBOURNE.

1887.

CONTENTS.

CHAPTER I.

JENNER'S EARLY TROUBLES WITH COW-POX INOCULATION.

CHAPTER II.

THE TRUE PEDIGREE OF ENGLISH VACCINE.

CHAPTER III.

THE GREASE OF THE HORSE AS A SOURCE OF "VACCINE" LYMPH.

CHAPTER IV.

WHAT IS COW-POX?

CHAPTER V.

THE EFFECTS OF VACCINE INOCULATION IN THE FIRST REMOVES FROM THE COW.

CHAPTER IX.

THE INCREASING DEATH-RATE FROM INFANTINE SYPHILIS.

THE
NATURAL HISTORY OF COW-POX

AND

VACCINAL SYPHILIS.

CHAPTER I.

JENNER'S EARLY TROUBLES WITH COW-POX INOCULATION.

EVENTS that took place ninety years ago will probably reveal
to our present scrutiny some aspects that were not notice-
able to contemporaries. It may be thought that Jenner's
original inoculations with the cow-pox are long past the
stage of historical criticism, that they have been scrutinised
abundantly, and that the closest examination of them will
yield nothing new. Speaking only for myself, I have to
express surprise at the number of new impressions, or
corrections of traditional teaching, that have come to me
from a close study of Jenner's own writings, and of the
early Jennerian literature, at first hand. I am not less
surprised at the uncritical manner in which Jenner's
writings had been read by his biographer, Dr. Baron, from
whom we have taken much on trust; or at the language
of an authoritative writer * some thirty years ago, when he

---

* *Papers relating to the History and Practice of Vaccination.* Parliamentary blue-book, compiled, with a preface, by John Simon. London,
1857.

gave the advice "to study thoroughly that masterpiece of medical induction, and imitate the patience and caution and modesty with which Jenner laid the foundation of every statement he advanced." It may be that our standard is higher now; but I am bound to say that when I did study thoroughly Jenner's three essays on cow-pox inoculation, I seemed to find myself dealing with reasonings which were anything but masterly, and with a writer who was never precise when he could be vague, and was never straightforward when he could be secretive. Any one who cares to try it will find that he has to hunt high and low in Jenner's papers for particular matters of fact, or for links in the succession of events, such as would be stated explicitly in their proper place by writers of the present day who make some pretension to scientific method, and such as were actually recorded with business-like candour by Woodville in Jenner's own time, and by Bousquet, Estlin, and Ceely, when the movement for "going back to the cow" arose some forty years later.

Every one will be ready to allow something for the novelty of the scientific proof, as well as something for the off-hand or casual manner of a country doctor in Jenner's time and circumstances; and, if that were all, I should not think it necessary to recall the incidents of eighty or ninety years ago. I go back to them in this chapter with a special object. I may say at once that the intention is not to deal with the whole Jennerian novelty, including the validity of the proof, the manner of endorsement by the heads of the profession, and the substantial recognition by a Parliamentary Committee. In that comprehensive spirit the story of the rise of vaccination has been treated lately by a hostile writer (Mr. White) with much literary ability, and with a degree of scientific knowledge commendable in a layman. Jenner's part in the history of vaccination is reviewed here only in so far as his own writings bear upon the original sources of vaccine lymph, and upon the direct effects of that virus as manifested in the earliest recorded cases.

Jenner's first inoculation for cow-pox was done on the 14th of May, 1796, upon James Phipps, aged six years, with matter from a large vesicle on the hand of Sarah Nelmes, a dairymaid, who had acquired the disease in milking cows. Writing on July 29th, 1796, Jenner says : "I have at length accomplished what I have been so long waiting for, the passage of the vaccine virus from one human being to another." No stock of lymph, however, was raised from James Phipps. Although Jenner's biographer, Baron, in speaking of this historic case,* says that "the boy went through the disease apparently in a regular and satisfactory manner," he speaks loosely : he had overlooked Jenner's parenthesis about "subsequent eschars on the inoculated parts." †

The historic case of James Phipps was not used to start a stock of lymph with ; nor does it appear that the other opportunities which arose about the same time at one or more farms in Jenner's neighbourhood, ‡ both in the cows and in the milkers, had been embraced for the purpose of repeating

---

* *Life of Jenner*, i. 137.

† *Inquiry into the Causes and Effects of Variolæ Vaccinæ, a Disease known by the name of the Cow-pox*, p. 31. The references in the sequel are to the paging of the edition of 1800, which contains the three essays together.

‡ *Loc. cit.*, pp. 15 and 17.

the experiment. It was not until nearly two years after that Jenner renewed his attempt with cow-pox.

On the 16th of March, 1798, William Summers, aged five years, was inoculated with "matter from the nipples of a cow," and the progress of the pustule was "similar to that noticed in the case of James Phipps;" that is to say, there were "eschars" subsequent to the formation of the scab. From the boy Summers matter was taken on the twelfth day, and inoculated on William Pead, aged eight years, of whose arm a coloured plate is given (Plate III.) representing a vesicle with the brownish centre fallen in. From the boy Pead "several children and adults were inoculated," whose subsequent progress is very meagrely indicated : three of them had an "extensive erysipelatous inflammation. It seemed to arise from the state of the pustule, which spread out to about half the diameter of a sixpence [very moderate size]. . . . By the application of mercurial ointment to the inflamed parts the complaint subsided without giving much trouble." The only case specified in this series is that of Hannah Excell, aged seven years ; and she was probably the only one from whom lymph was taken. Plate IV. shows three vesicles (of unequal size) on her arm at the ninth day. From her lymph was inoculated on the 12th of April upon two infants and two young girls. In one of these the virus failed to "take " ;* in another the vesicle "scabbed quickly without

---

* This was Jenner's second son, Robert F. Jenner, aged eleven months, who figures in the well-known story in marble, issuing from the modern Italian school, of "Jenner Vaccinating his own Child." When the boy was really inoculated a few years later, on a sudden emergency at Cheltenham, it was not with cow-pox matter, but with the old-fashioned variolous matter. (See Baron's *Life of Jenner*, i. 147, and ii. 43, 152.)

any erysipelas;" and of the remaining two we read that "the virus on the arm was destroyed soon after it had produced a perceptible sickening;"* and again, of one of them (Mary James), that the "pustule in the arm was destroyed." From the girl Mary Pead, whose vesicle "scabbed quickly without any erysipelas," matter was taken to continue the succession with : it was inoculated on J. Barge, aged seven years, whose progress is described in merely general terms. The series goes no farther than the boy Barge.

However, from Hannah Excell, two removes farther back, Jenner had preserved some vaccine matter dried upon a quill. This he took with him to London, in June, 1798, and gave to Mr. Henry Cline, the well-known surgeon, to make trial of. The purpose with which Mr. Cline used the vaccine enables us to measure Jenner's own estimate of the course of the vaccine vesicle, so far as his experience of it in the accidental form on the milker's hand, as well as in the experimental form on the child's arm, had taught him at that early date. Mr. Cline intended to use the remains of the vesicle as a pea-issue, and with that view he made the inoculation over the hip of a boy with hip-joint disease. Jenner had left London before the experiment was tried, and the news of it reached him in a letter. Mr. Cline wrote, under date 2nd of August, 1798 : "The cow-pox experiment has succeeded admirably. . . . The inflammation arising from the insertion of the virus extended to about four inches in diameter, and then gradually subsided without having been attended with pain or other inconvenience. The ulcer was not large enough to contain a

* *Further Observations, ed. cit.,* p. 105.

pea, therefore I have not converted it into an issue as I intended." That was what Cline wrote in the letter as Baron reproduces it from Jenner's papers ;* and the definite article in "*the* ulcer " proves clearly enough that Jenner had led Cline to *expect* an open sore as the sequel of the vesicle. However, when the time came for Jenner to publish Cline's letter (in his second pamphlet) as a valuable testimonial, the aspect of things had changed somewhat. Woodville had meanwhile succeeded in raising a London stock of vaccine, which did not lead to ulceration, although there was some trouble from eruptions in his patients, most of them the ordinary variolous eruptions from concurrent small-pox infection caught at the old Inoculation Hospital. Accordingly Jenner took the liberty of leaving out the sentence beginning, "The ulcer was not large enough," and of inserting instead of it, "There were no eruptions." *O tempora, O mores !*

Cline inoculated three children with matter from the boy's hip, but failed to raise vesicles; and on the 18th of August he wrote to Berkeley for fresh lymph, of which Jenner had none to send. It will be remembered that Jenner had carried his series two removes beyond Hannah Excell (who furnished Cline's vaccine) before he stopped it or failed to continue it. It is not upon himself, however, but upon Cline that he lays the blame of the stock having failed. Writing on the 27th of September, 1798, to Pearson, who was urging him to start vaccination in real earnest, he says : " It is painful to me to tell you that I have not an atom of the matter that I can depend upon for continuing the experiments. Mr. ——, when he inoculated the boy, did not take

* *Life of Jenner*, i. 152.

matter early enough to secure its efficacy." * It has to be said, however, that previous to the date of that letter, and shortly after his return from London, Jenner had made a third attempt to start a continuous series of cases, and to secure a permanent stock of lymph.

In *Further Observations* (p. 111) he says, " I have often been foiled in my endeavours to communicate the cow-pox by inoculation. . . . Four or five servants were inoculated at a farm contiguous to this place last summer [published in April, 1799] with matter just taken from an infected cow. A little inflammation appeared on all their arms, but died away without producing a pustule ; yet all these servants caught the disease within a month afterwards from milking the infected cows, and some of them had it severely." (He mentions the same failure, " Three or four servants at a farm were carefully inoculated with matter fresh from a cow," in a letter to Woodville in the end of January, 1799.) For whatever reason, Jenner does not appear to have inoculated from the cow-pox which these very servants acquired accidentally a month afterwards, although he was in want of lymph to send to his correspondents.

On November 8th Pearson again writes to Jenner : " If I can but get *matter*, I am much mistaken if I do not make you live for ever ; " and five days later (November 13th) he writes him more urgently than before to provide matter for distribution : " I wish you could secure for me matter for inoculation, because, depend upon it, etc. . . . By way of *se defendendo*, we must inoculate." Thus urged in all good faith by Pearson, Jenner must have felt as if

---

* The letter is printed by Pearson in his *Inquiry concerning the History of the Cow-pox.* London, 1798.

challenged to go on. Accordingly, we find that on November 26th he took matter from a poxed cow at a farm in the village of Stonehouse, where the disease had been passing successively from one cow to another since Michaelmas.* The results of that new trial, the fourth that he is known to have made with cow-pox matter, are not recorded in their completeness : they have to be collected from various scattered references. When Woodville wrote to Jenner on the 25th of January following, to tell him of his own success in establishing a stock of vaccine from a cow in London, Jenner replied by return of post,† and thus adverts to a late attempt of his, which can have been none other than that with the Stonehouse cow-pox, on November 26th : " Whether to the cold season of the year, or to what other cause it can be ascribed, I know not, but out of six patients that I lately inoculated, two of them only were infected." Of the four uninfected, two (on November 27th) are known‡ to have been the children of Mr. Hicks, of Eastington, who were successfully vaccinated for the first time with Woodville's own lymph in the end of February following.§ The two who were infected are introduced with a good deal of detail in *Further Observations*,‖ in order to illustrate

* *Further Observations, ed. cit.* p. 99.

† Baron, i. 308.

‡ Baron, i. 303. The biographer does not seem to have known that these vaccinations failed (although he mentions "two children of Mr. Hicks" as having been vaccinated with Woodville's lymph some three months after). He specially calls attention to their "vaccination" on the 27th of November, "to disprove an assertion subsequently made that the first vaccination performed by Dr. Jenner after the publication of his *Inquiry* [in June, 1798] was with virus furnished by Dr. Pearson."

§ Baron, i. 324.

‖ *Ed. cit.*, pp. 99—103.

Jenner's new opportunist doctrine of primary and secondary, or essential and unessential effects of cow-pox inoculation.

Susan Phipps, aged seven years, was inoculated on December 2nd with matter which had been taken from the cow on November 26th, and dried on a quill : the vesicle did badly, became a phagedenic ulcer, the size of a shilling, and healed by granulations after several weeks. From her matter was taken on the eleventh day by "Mr. D., a neighbouring surgeon," and probably also by Jenner himself ; at all events, Mary Hearn, aged twelve years, was inoculated with matter taken from the arm of Susan Phipps, and in her case also " the progress of ulceration " had to be checked about the end of the second week and beginning of the third, by repeated applications of mercurial ointment.

There is nothing known of any stock having been raised with the matter from Susan Phipps, taken by " Mr. D., a neighbouring surgeon," nor does Jenner say that he continued the succession beyond Mary Hearn (the second remove from the cow). In fact, he was again stopped, either by the fear of ulceration, which, as we shall see, had been secretly haunting him all through his experiments, or by some other obscure cause. It was not until he received, on the 15th of February, 1799, a thread soaked in lymph from Woodville's continuous series of vaccinations in London, that Jenner was able to establish a supply at Berkeley. He inoculated his nephew, Stephen Jenner, aged three and a half years, with one half of the thread, and James Hill, aged four years, with the other half, both of whom were infected satisfactorily ;[*] and from James

----

[*] *Further Observations, ed. cit.,* pp. 130—132.

B

Hill he took matter to Eastington, where he inoculated the two children of Mr. Hicks, and sixteen others in that gentleman's factory or house. Jenner had at length got a lymph to his mind, although it was not of his own breeding. Writing on the 13th of March to Pearson, he describes the effects of the London lymph as follows :* "The character of the arm is just that of cow-pox, except that I do not see the disposition in the pustules to ulcerate, as in some of the former cases." Pearson had by that time issued more than two hundred threads of vaccine matter to practitioners all over the country, as well as abroad, along with a circular letter of directions; so that Jenner was now only one of many who were vaccinating continuously with the London lymph.

His nephew wrote to him from London to come up and wear the laurels which he had won, and to prevent others from wearing them; and accordingly Jenner went to the capital in the end of March, leaving the vaccinations to be carried on in his absence mainly by Marshall, a practitioner at Eastington. The situation was somewhat difficult, and it cannot be said that Jenner comes out of it as creditably as we could have wished. He seized upon the variolous eruptions that were complicating Woodville's cases at the Inoculation Hospital as a pretext for insinuating, in his correspondence and otherwise, that the lymph was not genuine; although he himself was using that very lymph, having absolutely failed to raise a stock of his own'! In a few weeks Woodville got over the eruptions difficulty; but Jenner kept up the cry for several years.† Meanwhile, he saw the opening for

* Baron, i. 316.

† See his letter of 4th of March, 1801, to Dr. Waterhouse, of Harvard.

a new stock of lymph of his own finding, which would be under no reproach as to eruptions. Thomas Tanner, a veterinary student from Gloucestershire, was the means of getting some cow-pox matter for him from Clarke's dairy in Kentish Town, " some time in April."* Although Jenner was at hand in London, he did not attempt to start a series of arm-to-arm inoculations with the new matter ; but sent Tanner with it at once to his friend Marshall, at Eastington, who was then busy vaccinating with Wood-ville's lymph. Marshall writes to Jenner on the 26th of April, giving an account of 107 vaccinations, with accurate details, which latter Jenner thought it superfluous to pub-lish. He says nothing about any Kentish Town lymph brought by Tanner, but ends his letter with a declaration which sounds rather odd in the circumstances : "In the cases alluded to here, you will observe that the removal from the original source of the matter [Woodville's Gray's Inn Lane cow] has made no alteration or change in the nature or appearance of the disease, and that it may be continued *ad infinitum* (I imagine) from one person to another with-out any necessity of recurring to the original matter of the cow." †

Meanwhile Robert Tanner, a resident in Gloucestershire, had also discovered a case of cow-pox, at North Nibley in that county, and sent matter from it to Jenner in London. All

U.S., printed by that author in his second pamphlet, p. 18. (Cambridge, U.S., 1802.)

* Jenner, in a letter to Ring, August 16th, 1799 (Baron's *Life*, i. 356). The characters of the disease in this cow are nowhere stated : an omission not to be excused considering the many forms of "spurious" pox in the cow.

† Jenner's *Continuation of Facts and Observations*, ed. cit., p. 155.

that we hear of this is that Jenner, on the 13th of April, gave a portion of it to Mr. Knight, a well-known court and army surgeon, who is not stated to have done anything with it.* In the end of June, Jenner went back to Gloucestershire, and doubtless received a conversational account from Marshall of his experiences with the Kentish Town lymph, which had been sent to him to establish a stock from in April. As Jenner, like most of his successors, had had serious trouble with matter direct from the cow's teats, it would have been interesting to him to know if his deputy had fared better than himself. Marshall, indeed, wrote another letter (which is printed by Jenner in his third pamphlet without date, but ought to be dated 8th of September, according to Baron †), wherein he speaks of his vaccination experience as a whole, adding in the most casual way, in a postscript, that 127 out of a total of 423 vaccinations (or just thirty per cent.) had been done from the independent stock of matter brought by T. Tanner from the Kentish Town dairy. On that peculiarly confused evidence, Jenner rested his formal claim ‡ to be the possessor of a stock of lymph more genuine than Woodville's, from which he professed to satisfy the many applicants who came to the discoverer himself for lymph.

The practical importance of that claim may be judged from one instance, not to multiply evidence on a disagreeable topic. In the quarrel which had now begun between Jenner and the discoverers of the London lymph, John Ring professed his distrust of the latter, and entered into

---

* Baron, i. 323.

† *Loc. cit.*, i. 325, note.

⁺ *A Continuation of Facts and Observations*, ed. cit., p. 162.

correspondence with Jenner, with the result of becoming the most active of his London supporters. More than once he writes to Gloucestershire for genuine lymph ; and on the 18th of September, 1799, Jenner sends him some, with this certificate, that "it had been passing from one patient to another for upwards of six months," and that it was "from the source mentioned at the conclusion of my second pamphlet." Whether Ring took the trouble to refer to the book in order to discover what that source really was, does not appear ; but, in point of fact, it was no other than Woodville's own London lymph.*

The truth is that Woodville's discovery of a benign lymph, or a lymph which had not produced ulcerating vesicles, was a welcome relief to Jenner from his own well-grounded fears. He knew much more of the real nature of cow-pox than the London vaccinators, although he was at the same time much less experienced than they were in all that related to the "management" of lymph, or to the technical art of inoculating. Having once got over the initial difficulties by the fortunate aid of others, he had no wish to go back to the cow ;† at least, not with his own hands, or for his own purposes. Apart from Marshall's unauthenticated experiments made when Jenner was in London, it does not

---

* I have collected a good deal more of that sort of circumstantial evidence on the trumped-up rivalry of the lymphs ; but I forbear from enlarging on an incident in the early history of vaccination which Woodville and Pearson, to their credit, made no attempt to magnify the importance of at the time.

† "There is, therefore, every reason to expect that its effects will remain unaltered, and that we shall not be under the necessity of seeking fresh supplies from the cow."—Jenner, under date December, 1799, *id. cit.*, p. 162.

appear that the latter ever sought to go back to the cow. His experience of the cow had not been reassuring ; and his last attempt, on the 2nd of December, 1798, had been the most discouraging of all in respect to the phagedenic ulceration of the vaccinated arms.   It was a precisely similar experience of "enormous pustules, violent inflammation, and slow-healing ulcerations after the fall of the crusts," following on the inoculation of an infant with primary lymph from the Passy cow, in 1836, that led Bousquet to say : "C'est de ce moment que j'ai compris pour la première fois les frayeurs de Jenner."* It is even permissible to speculate that, but for Woodville's fortunate hit, which nothing in the whole experience of cow-poxing has ever equalled, Jenner might have been absolutely deterred from pursuing his project, the more so that he did not at that time look to it to supersede variolous inoculation except under the particular circumstances which he specifies at the end of his first essay.   (*Vide infra* p. 40.)

I shall have to speak briefly, in chapter iii., of the lymph raised by Jenner from horse-grease, and actually put into circulation by him as a practical assertion of his whimsical theory, and of his own originality.   Meanwhile it will be necessary to take the pedigree of Woodville's and Pearson's cow-lymph.   And, as we are here at the very fount and origin of the world's vaccine, a special attention is due to the facts and circumstances.

* *Sur le Cow-pox découvert à Passy (près Paris)*, 1836.   Paris, 1836, p. 21.

CHAPTER II.

THE TRUE PEDIGREE OF ENGLISH VACCINE.

On Sunday, the 20th of January, 1799, word was brought to Dr. Woodville, physician to the Inoculation Hospital, that the cow-pox had appeared in the cows at a dairy in Gray's Inn Lane.* The next day he went to see the disease, taking with him Tanner, the veterinary student from Gloucestershire, who was supposed to know something of it. Three or four cows were found to be affected with "pustulous sores on their teats and udders." There were about two hundred cows in all, and of these four-fifths became eventually infected, those not in milk escaping.

Woodville thus narrates the event to Jenner in a letter four days after :† "As he [Tanner] declared it to be the genuine disease, I that day inoculated six persons with the matter that he procured from a cow which appeared to be the most severely affected with the pustular complaint. On Wednesday I again called at the cow-house to make further inquiries, when I was much pleased to find two or three of the milkers were infected with the disease, one of whom exhibited a more beautiful specimen of the disease than that which you have represented in the first plate [large bluish-white vesicle on the hand, with centre fallen in]. From

---

* *Reports of a Series of Inoculations for the Variolæ Vaccinæ or Cow-pox.* By William Woodville, M.D., Physician to the Small-pox and Inoculation Hospitals. London, 1799.

† Baron's *Life*, i. 307.

this person I charged a lancet with the matter, which appeared different from that taken from the cow, as that of the former was purely lymphatic, and the latter of a purulent form. With this lymphatic matter I immediately inoculated two men at the hospital. Finding now [*i.e.* after the infection of the milkers] there could be no doubt of the disease," he went to inform Sir Joseph Banks, Pearson, Willan, and others, who visited the cow-house along with Lord Summerville and several more on the day following. Jenner's book was produced, and the appearance in Plate I. compared with the vesicles on the hand and arm of one of the milkers, and pronounced to be very similar. Matter was a second time taken from the milker's hand or arm, Woodville proceeding straight to the Inoculation Hospital with it, and there inoculating six more (making fourteen direct inoculations in all).

To this communication from Woodville, dated 25th of January, Jenner replied by return of post. He wishes, he says, that he could be at Woodville's elbow. "After the description you have given, there can be no doubt, I think, that the disease is the true and not a species of the spurious cow-pox. In the account of the appearance on the milker's hand, the report of my friend Tanner merits great confidence." In view of the strong position that Jenner had taken up on the question of spurious cow-pox only six months before, and of what he said about it again a month or two later, he was certainly not over-critical, from his own point of view, in giving a warranty of genuineness to Woodville's lymph. This curious piece of sophistry touching the "genuine" and the "spurious" in cow-pox will be referred to later on; meanwhile the Gray's Inn Lane cow became, as a matter of fact, the source of English vaccine.

Woodville has reported in detail the circumstances and results of his inoculations (200) from the 21st of January down to the 18th of March, and has given a tabular pedigree of the lymph used in about 250 cases subsequent to these. Thus we are enabled to trace to its source Jenner's own stock of lymph, which was sent to him on the 15th of February, and with which he started his own first continuous series. It was taken from Ann Bumpus, aged twenty, who was inoculated on the 6th of February from Sarah Butcher, a healthy girl, aged thirteen, whose vaccinifer, on the 30th of January, was Jane Collingridge, a healthy, active girl, aged seventeen, she herself being one of the first group inoculated with purulent matter direct from the cow's teat on the 21st of January. The vaccine vesicle on Collingridge's arm on the eighth day was perfectly circular, and of a lemon-coloured tint; on the eleventh day it was inflamed at the margin, and beset with minute confluent pustules. (In one of Jenner's first pair of vaccinations from this stock a month later, that of James Hill, the vesicle on the eleventh day was not merely "surrounded by an inflammatory redness the size of a shilling," but that area was also "studded over with minute vesicles.") On the thirteenth day the vesicle was scabbing, but on the seventeenth day the scab was "in a state of suppuration," which appears to have ended in cicatrisation without further incident. Collingridge was one of those who were variolated a few days after being vaccinated, either in anticipation of Jenner's advice, or in compliance therewith, the latter having written to Woodville: "I answer your letter by return of post to suggest (what perhaps is needless) the immediate propriety of inoculating those who may resist the action of the cow-pox matter, and may have been exposed to variolous contagion at the hospital." * This practice of variolation was carried out in a number of the earlier cases where the development of the vaccine vesicle was thought problematical, the same having failed altogether in three out of the fourteen original cases. The variolous inoculation in Collingridge ran its course (producing 170 pustules) independently of the vaccine vesicle, which was in like manner unaffected by the concurrence of variola. The artificial inoculation of small-pox accounts for a good deal of the eruptions in Woodville's early practice ; small-pox caught naturally in the atmosphere of the hospital (where variolation was the regular business at the time) accounts for a good deal more, the two classes of applicants having mixed together indiscriminately ; while some small part of the eruptions (not variolous in type) was almost certainly the effect of the cow-pox matter itself. Although Jenner made much capital out of these eruptions, by way of getting preference for his own stock of lymph, it has to be remembered, not only that his own lymph came from Woodville's hospital, but that it was Jenner himself who advised Woodville to adopt concurrent variolation in the first experiments.

Returning to the pedigree of Jenner's stock, Sarah Butcher, aged thirteen, who was the second remove from the cow, had also on the eleventh day "suppuration at the inner edges of the tumour, redness at the outer edge very extensive ;" on the sixteenth day the vaccinated area had

* Baron's *Life*, i. 308.

scabbed at the centre. She was variolated the same day with small-pox matter, which merely produced a little redness at the spot. From her on the seventh day was vaccinated Ann Bumpus, aged twenty, the direct vaccinifer of Jenner's first cases in his unbroken series. She seems to have had a regular vaccine vesicle, and on the eighth to tenth days to have had the premonitory lassitude and rigors of an attack of small-pox caught from aërial contagion. She had 310 variolous pustules; they had dried up by the twenty-second day (from vaccination), at which date she was artificially inoculated with small-pox without effect. The matter taken from Bumpus was despatched to Jenner on the 15th of February, being the seventh day; the five vaccinations performed at the hospital with lymph from the same vaccinifer were not done until the 18th of February, and of these only one (Sarah Dixon, aged nineteen) had a crop of pustules (174), having shown the malaise and rigors of commencing natural small-pox on the tenth and eleventh days, the eruption beginning to come out two or three days later. All the others had the ordinary vaccine vesicle, and no eruptions, unless that "several pustules appeared on the eleventh day at the margin of the tumour" on the arm of James Cummins, aged fourteen weeks. All the five were variolated soon after without effect.

In tracing the antecedents and collaterals of Jenner's lymph, we have given a fair sample of Woodville's practice; and it is unnecessary to follow the details further. On one important point, namely, Jenner's old *bête noire* of spreading ulceration, Woodville's experience was singularly fortunate: "We have been told that the cow-pox tumour has frequently

produced erysipelatous inflammation and phagedenic ulceration, but the inoculated part has not ulcerated in any of the
cases which have come under my care, nor have I observed
inflammation to occasion any inconvenience, except in one
instance, where it was soon subdued by the application
of *aqua lithargyri acetati*. It would seem, then, that the
advantages to be derived from substituting the cow-pox for
the small-pox must be directly in proportion to the greater
mildness of the former disease."

Two points remain to be noticed before we leave
Woodville's practice. The first is that the strain of lymph
from the dairymaid's hand, which was started alongside the
more purulent matter direct from the cow's teat, was allowed
to lapse at the third remove, the reason not being apparent
from the details (which are meagre), while nothing is said in
the commentary. The second point is that the dairymaid's
lymph at the first remove (James Crouch, aged seven) was
inoculated back upon a healthy cow's teat at the Veterinary
College. We are simply told that this operation "produced
the disease in the cow," and that "a man-servant, by milking
this cow, was also affected with an extensive tumour upon
his thumb; this soon acquired a livid blue colour, and was
attended with a considerable degree of fever, and with
a rash upon his ankles and feet." * The matter for the
retro-vaccination of the cow seems to have been taken from
James Crouch about the 4th or 5th of February, or perhaps
earlier, and the cow's lymph was inoculated on the 18th of
February (allowing about twelve days for the development
in that animal) upon three grown-up persons, one of whom
had erysipelas, and all of them a moderate amount of

------

* Woodville, *loc. cit.*, under the heading of the thirty-ninth case.

eruption. Each of the three became the vaccinifer of numerous others, and that strain plays as large a part as any in the subsequent vaccinations at the hospital.

Besides Woodville's stock of lymph, we have to notice in passing the claim made by Pearson* to have introduced a parallel and contemporaneous strain of vaccine matter from another London dairy. After referring to the essay which he published in 1798, in support of Jenner's theoretical position, he proceeds as follows : "Vaccine matter was in vain inquired for, and Dr. Jenner had discontinued the inoculation about the time of publishing his book above mentioned. But from the curiosity excited by my inquiries among the milk-farmers near London, as appears from the *Inquiry into the Cow-pox* which I published, but principally owing to the attention of Dr. Woodville, information was communicated in January, 1799, that the cow-pox was epizoötic in Gray's Inn Lane ; and at the same time I received the agreeable intelligence that this disease was also raging in the largest stock of cows on the New Road, near Paddington, to which no one could gain admittance but myself. With vaccine matter procured from these sources, Dr. Woodville instituted the trials of the new inoculation in the Small-pox Hospital; and I carried on mine in certain situations instead of the small-pox, and among such persons as I induced to undergo the experiment ; besides, we promoted the practice by furnishing Dr. Jenner, of Berkeley, and other practitioners, with London vaccine matter for the repetition of the cow-pox inoculation in Gloucestershire and other places." He proceeds to say that they had no

---

* *An Examination of the Report*, etc. By George Pearson, M.D., F.R.S., Senior Physician to St. George's Hospital (London, 1802), p. 43.

occasion, after the first series of vaccinations, to recur to the cow for fresh matter. In Pearson's former essay on the cow-pox, he refers to the milk-farm near the New Road, Marylebone. It contained from eight hundred to a thousand milch cows; and he was told on inquiry that cow-pox was pretty frequent there, especially in winter, being supposed to be due to sudden change from poor to rich food. Three of the men about the place had formerly caught the disease from cows, and bore the scars of it. I have not found any precise description of Pearson's New Road cow-pox, or any authentic narrative, comparable to that of Woodville, of the first vaccinations therewith. I am inclined to think, from the studied vagueness of his language, that he confined himself, in his vaccination practice, to the stock from the Gray's Inn Lane dairy.

CHAPTER III.

THE GREASE OF THE HORSE AS A SOURCE OF "VACCINE" LYMPH.

If Jenner had been able to give practical effect to his sophistically adopted theoretical fancies, no lymph that was not derived, either directly or indirectly, from horse-grease, would ever have been used for Jennerian vaccination. The only "genuine" cow-pox, in his estimation, was that which was conveyed to the cow's teats by the hands of men who had been dressing the sore heels of horses;* and, in his

---

* *Inquiry*, p. 7; *Further Observations*, p. 90.

zeal for that definition of genuine pox in the cow, he excommunicated the "spontaneous" cow-pox, a sporadic malady mostly of the spring season and a rather rare liability of heifers in their first milk. Practical men gave little heed to Jenner's fancy for horse-grease, not knowing the logical need for it ;* but he himself went back to his original doctrine after the public had accepted cow-pox, and notwithstanding that the "Jennerian lymph" of practice was Woodville's lymph, and in nowise connected with horse-grease.†

That the grease of the horse's hocks produced vesicles, and afterwards sores, on the hands of blacksmiths, farriers, and stablemen, was or is admitted by all authorities. It is admitted also that the vesicles and sores so produced are not unlike those caught from the cow's teats, the vesicles enlarging and bulging at the periphery, degenerating into phagedenic ulcerations, apt to be attended with swelling of the nearest packet of lymphatic glands and with much constitutional fever and even delirium. Jenner mentions several such cases, and at p. 96 gives the details of one case reported by Fewster :

"On the middle joint of the thumb of the right hand there was a small phagedenic ulcer, about the size of a large pea, discharging an ichorous fluid. On the middle finger of the same hand there was. another ulcer of a similar kind. These sores were of a circular form, and he [the patient] described their first appearance as being somewhat like blisters arising from a burn. He complained of excessive pain, which extended up his arm into the axilla. These symptoms and appearances of the sores were so exactly like the cow-pox, that I pronounced he had the distemper from milking cows. He assured me that he had not milked a cow for more than half a year, and that his master's cows had nothing the matter with them." His master,

* See note on p. 73.   † Baron's *Life*, ii. 226.

however, had a greasy horse, whose heels the patient had dressed twice a day for the last three weeks or more ; and it was remarked that the smell of his hands was much like that of the horse's heels.

Three cases of these horse-sores on men's hands occurred together in Jenner's own practice* in February and March, 1798 ; one of the men appears to have conveyed the disease to the cows in milking them : "their nipples became sore in the usual way with bluish pustules ; but, as remedies were early applied, they did not ulcerate to any extent." From the sore or "pustule" on the hand of another of the three men, Jenner took matter, and on the 16th of March, 1798, inoculated John Baker, aged five. It requires a good deal of research to find out all that happened to John Baker. We are told in the first narrative that he became ill on the sixth day with symptoms similar to those excited by cow-pox matter, and that, on the eighth day, he was "free from indisposition." Then follows a plate, showing the pustule, an enormous yellowish vesicle with a tumid periphery, a broad central area of brownish sloughing cuticle, and an angry blush of the skin around. Next page we read that the variolous test could not be applied to the child, because he "felt the effects of a contagious fever in a workhouse soon after the experiment was made." Another item of information about John Baker was made public a year after in a note to the second pamphlet (*ed. cit.,* p. 93), by which we learn that the words, "felt the effects of a fever," meant that the boy died of it. But the greatest light was thrown upon this case of horse-grease inoculation after Jenner's death, when the ingenuous Baron printed

* *Inquiry,* p. 32.

a number of the discoverer's papers and memoranda just as he found them. In his second pamphlet (*ed. cit.*, pp. 91–92) Jenner had stated a number of reasons for concluding that cow-pox is, in fact, the bovine communicated form of horse-grease, the last reason being "the progress and general appearance" of John Baker's "pustule." In the copy from which Baron reproduced the enumeration of reasons,* we find "the disposition of the pustule to run into an ulcer" alleged in further proof of the similarity of Baker's infection to the vaccine disease, as indeed it well might be. From all these veiled and scattered references we may conclude that Jenner's first case of horse-grease inoculation produced an ulcer, which was doubtless of the same inveterate and specific type as in the accidental disease on the hands of stablemen and farriers; also that the boy died in the work-house soon after the experiment; but whether the "contagious fever" of which he died was simply the result, direct or indirect, of his poisoned arm, we can form no opinion of our own, owing to the arbitrariness (or "caution," as an apologist would name it) of the Jennerian records.

Various attempts were made to prove by experiment Jenner's fanciful doctrine that horse-grease was the original of cow-pox. Sacco of Milan, and Loy of Whitby, were the most systematic experimenters. There can be no doubt that the inoculation of the human arm (as well as the cow's udder) with the pungent matter from the horse's greasy hocks has been followed by a bleb, or pock, or vesication which could not be distinguished from the vesicle of cow-pox inoculation. In both cases the vesicle grows at its periphery by eating away the margin of tissue under

* *Life of Jenner*, i. 248.

c

cover of the skin, encroaching to the depth as well as to the breadth, and so producing the characteristic fulness and distension at the edge which is indicative of an active corroding process, all the while that the centre is already sloughing. The same subcutaneous or subcrustaceous corroding tendency is admitted by Jenner himself, at the end of his third pamphlet, as being the exception rather than the rule of the pathological process. He speaks of vaccine vesicles with "a creeping scab of loose texture, and subsequently the formation of limpid fluid at its edges." In a letter to Waterhouse, 4th of March, 1801, he refers to the creeping scab as a common form of miscarriage in vaccination: "That which appears next in frequency is, according to my observation, a soft scab gradually creeping around the punctured part until it has attained the size of a sixpence, or a larger size, and then dying away, instead of a hard red spot converting itself in four or five days into a vesicle." [*]

The puffed margin of the vesicle marks an order of progress common to various specific ulcerations when inoculated on the human skin, and to the venereal among the rest. According to Ricord, the chancrous pustule, following the experimental inoculation of matter on the skin from a primary syphilitic sore, has characters which might easily lead to its being mistaken for a vaccinal vesicle.[†] In Ricord's [‡] plates of experimental chancres, we see large

---

[*] Waterhouse, ii. p. 112.

[†] Quoted by Diday, *Traité de la Syphilis des Nouveau-nés et des Enfants à la mamelle.* English edition, 1859, p. 54.

[‡] *Traité complet des Maladies Vénériennes.* Paris, 1851, Plate I., Figs. 6 and 7 ; and Plate III., Figs. 7, 8, and 9.

whitish vesicles or pustules, and examples of what might be called a "creeping scab." In Fig. 7 of his third plate the dried-up or gangrenous condition of the centre, and the advancing fulness around the margin under cover of the unbroken skin, are just as characteristic as if the figure had been one showing the inoculated cow-pox. Mr. Henry Lee's plates of the experimental inoculation of venereal sores bear out the same conclusion.* More particularly Fig. 2 of his second plate, representing the pustule at the eighth day after inoculation, with matter once removed from a serpiginous sore, has the depressed centre and the puffed vesicular whitish ring round it, in as typical a form as any vaccinal vesicle has (for example, that of Hannah Excell in Jenner's fourth plate). At a later stage this umbilicated vesicle broke; and the figure underneath it in tho plate shows the "exceedingly irritable" ulcer that ensued.

It is not surprising, therefore, that the grease of the horse's hocks should have passed through a vesicular stage, like that of the cow-pox, when inoculated on the human skin. But, in regard to its further and complete natural history, the records of Loy's experiments are of no value.† He follows the example of Jenner, not merely in the matter of handsome type and wide margins, but also in the arbitrariness of his statements, and in the withholding of every detail which would enable an independent reader to form an opinion of his own. His authority may be accepted when he tells us that primary inoculations of man

* "Syphilitic Inoculation," in *Med. - Chir. Trans.,* xliv. (1861), p. 238.

† John G. Loy's *Account of Experiments on the Origin of the Cow-pox.* Whitby, 1801.

with the matter of horse-grease produced vesicles like the
vaccine vesicles ; but, as a scientific record of experiments
from first to last, his book is not up to the average standard
of precision, and is defective more particularly, in that
it never mentions the after-history of the vesicle.

A few months before Loy's experiments Jenner himself
had sent to the Inoculation Hospital in London a stock of
lymph which was raised on the cow's teat by his veterinary
neighbour, R. Tanner. It is clear, however, from the
evidence, that cow-pox matter had been inoculated on the
teat a few days before the horse-grease was applied to the
same. Long after, on July 23rd, 1813, we find a note by
Jenner, referring to "equine virus which I have been
using from arm to arm for these two months past, without
observing the smallest deviation in the progress and appearance of the pustules from those produced by vaccine."
Again on the 17th of May, 1817, there is this memorandum :
"Took matter from Jane King (equine direct) for the
National Vaccine Establishment. The pustules beautifully
correct." *

From the fact that the matter of horse-grease could
produce correct vesicles on the human arm, loose logic has
constructed the theory that horse-grease is horse-pox, or
variolæ equinæ. First of all, a distinction was made by
Loy between true grease, which had constitutional disturbance, and spurious grease, which was purely local : the
distinction was, of course, an arbitrary one, and made to
suit that experimenter's convenience. Next, it was made
out that the grease was a composite disease, and that two
morbid processes co-existed in it, namely, " horse-pox " and

* Baron, ii. 226.

"the grease" as vulgarly understood among farmers and others. Lastly, the horse-pox, so distinguished, was named *variolæ equinæ* on the analogy of Jenner's equally fanciful designation of the pap-pox of the cow as *variolæ vaccinæ*. It then remained to give a systematic account of *variolæ equinæ* in keeping with these developments of doctrine : such an account as may be read in some veterinary text-books, or in Seaton's *Handbook of Vaccination.* In the latter it is stated (p. 78) : "They [the vesicles of horse-pox] have absolutely the same structure as the vaccine or variolous vesicle, and yield, though generally in small quantity, a viscid and slightly yellowish lymph. By the ninth, tenth, or eleventh day, many of them burst, exuding, often copiously, a viscid serous or sero-purulent fluid : incrustation going on progressively, and forming scabs or crusts, which from the fifteenth to the twenty-fifth day detach themselves, leaving whitish superficial cicatrices." Be it noted that the structure is the same as that of the vaccine *or variolous* vesicle.

Let us leave these ideal or simplified pictures, and turn to the concrete realities of horse-grease as Jenner knew it, and as it still occurs in ordinary country experience. Academical subtleties apart, there is no ambiguity about horse-grease ; and I think that I am myself familiar enough with its appearance to be able to make a correct diagnosis.

I take the following account from Professor Hering, of the Veterinary College at Stuttgart : *

"The so-called 'acute grease' consists in an erysipelas of the skin of the horse's hocks, which not unfrequently extends down the posterior surface of the cannon bone (metatarsus or metacarpus) ; it gives rise

* *Ueber Kuhpocken an Kühen.* Stuttgart, 1839.

at first to small vesicles (mostly overlooked among the hair), which burst and discharge an acrid lymph with a peculiar odour; later on, these vesicles, owing to the parts being wetted or covered with filth, or badly treated, are apt to pass into herpes-like chaps (chronic grease), which are slow to heal, and at length produce various degenerations of the skin and of the underlying tissues. In other cases the inflammation, being very circumscribed, extends more to the depth; and pieces of skin, more or less considerable in extent, are exfoliated as if from a dry gangrene. Viborg saw reddish warts arise, which Sacco would call small condylomata. The disease is more common in damp and low-lying places, in wet seasons, and in common breeds of horses, than in dry and elevated places and among the better breeds. The first or acute part of the disease is for the most part unnoticed, or treated with domestic remedies, and on that account comes before veterinary surgeons much less often than its sequelæ.

"Meanwhile, we are just as little in a position to deny the fact that the grease produces in the human subject and in the cow a disease like the cow-pox, as we are unable for the present to find any explanation of the fact itself *in the slight similarity between the one disease and the other.*"

And that, I conceive, will be the conclusion come to by every one who has no particular interest in constructing a consistent body of doctrine from the point of view of human small-pox. For the rest, I shall merely recur here to the characters of the disease in man, when it is accidentally caught from the horse's greasy hocks—characters about which all observers are agreed. Jenner thus states them:

Case of Thomas Pearce: "Sores on the fingers, which suppurated, and which occasioned a pretty severe indisposition." Case of Abraham Riddiford: "Very painful sores in both his hands, tumours in each axilla, and severe general indisposition." Case of William Morris (communicated by Fewster): "Two small phagedenic ulcers on the fingers, about the size of a large pea; excessive pain extending up the arm to the axilla; three days after, still complaining of pain in both his hands, nor were his febrile symptoms at all relieved; the ulcers had now spread to the size of a seven-shilling gold coin; and the ulcer, which I had not noticed before, on the left forefinger equally painful. Escharotics applied; got well in something more than a fortnight; lost the nails from the thumb and fingers that were ulcerated." *

* As I am concerned here only with horse-grease as an actual and historical source of lymph for inoculation, it is not necessary to enter upon the theoretical question whether the vesicular eruption of certain parts of the horse's skin (especially around the mouth), which was described by Lafosse in 1860, and by Bouley in 1863, and named by the latter "maladie pustuleuse vaccinogene," is the same as the grease of Jenner's practice. An account of it, and of the inoculation experiments, is given by Fleming in his series of papers on "Human and Animal Variolæ," in the *Lancet*, 1880, vols. i. and ii. Another inquiry of the greatest interest for this question is detailed by Dr. E. Klein in the *Report of the Med. Off. Loc. Gov. Board* for 1880. (Appendix, p. 183.) The disease appeared in two horses which had just come to London from the country. It began as red pimples of the skin at the angles of the mouth, which became vesicles, and then ulcers, leaving "papular" scars; the ulceration also existed on the mucous membrane; but there was no disease at the usual situation of grease, namely, the heels. In the accidental inoculation of a stableman, and in numerous experiments on animals, the vesicles that were produced broke, and left ulcers, sometimes deep, generally of the eating and indurating kind, and attended with much constitutional disturbance. Dr. Klein was disposed to infer that "the induration of the basis and margins of the ulcers in our experiments" served, along with other points, to distinguish the disease in these two horses equally from "horse-pox" and from cow-pox. But the phagedena (which he had previously mentioned), and (or) the induration, are characteristic of various ulcerous processes, when these are communicated by contact or by inoculation from subject to subject; they are characteristic of venereal ulceration, of cow-pox ulceration, of horse-grease ulceration, and, perhaps, of still other types of inveterate soreness. In the two horses in question, the process was localised around the mouth, and was absent from the hocks. It is quite conceivable that the animals had caught it on their mouths, either from direct contact with an ordinary case of greasy hocks, or indirectly by contact of their mouths with matter from the same.

It will be convenient to point out, in this connection, that the polemic of Squirrell against vaccination when it was first started * was based upon the assumption that the grease of the horse's hocks was the real source of vaccine matter, such being Jenner's theory, but not his practice. Squirrell's arguments were therefore rather easily met by practical vaccinators; but on general grounds they are still worth quoting : "On reading Dr. Jenner's account of the origin of the cow-pox [from horse-grease], I was struck with such horror and aversion, that I could not, as a man of honour or feeling, submit to or coincide with vaccination. . . What in the name of God could have induced him to have introduced a disease of so filthy a nature, and apparently, according to his own account, such a dangerous tendency ? I should have imagined that his own description would have furnished him with the most powerful argument against it." † Squirrell does not seem to have known how hesitating and full of fears Jenner was, owing to his uniformly discouraging experience of ulceration, following not only the casual but also the experimental insertion of cow-pox and horse-grease matter. He overlooks also the modest *rôle* assigned to vaccination in the original *Inquiry.* ‡ Moseley had a more

---

* *Observations on the Cow-Pox, shewing that it originates in Scrophula, and is no security against the Small-Pox.* London, 1805.

† *Loc. cit.,* p. 4.

‡ "Should it be asked whether this investigation is a matter of mere curiosity, or whether it tends to any beneficial purpose, I should answer that, notwithstanding the happy effects of inoculation, etc. . . I have never seen fatal effects arise from the cow-pox, even when impressed in the most unfavourable manner, producing extensive inflammations and suppurations on the hands; and as it clearly appears that the disease leaves the constitution in a state of perfect security from the infection of

correct, and therefore a more generous appreciation of the course of events than had Squirrell; he saw quite clearly that Jenner's hand had been forced by Woodville's and Pearson's success with a benign lymph, while he himself was still trying to deal with the problem of ulceration, with or without phagedena, as a sequel of the vesicle. "Unfortunately for society," says Moseley, "for Dr. Jenner, and the credit of his discovery, he was not left to prosecute it deliberately in the country, and to investigate it in a quiet philosophic manner, through a succession of many experimental years. The manufacture was still in embryo, when the raw materials were brought, unfit for use, to market; and they were snatched from his possession, in their crude state, by a set of medical jugglers, besotted and stupefied with the gigantic novelty, and scattered like firebrands among the Philistines." *

It is necessary, however, to add to this that Jenner, whenever he heard of Woodville's success, not only adopted the vaccine, but came up to London and began that advocacy of his claims which resulted in Admiral Berkeley's parliamentary committee in 1802, and in the grants of public money, and all other acts of public recognition in the years following. It is not surprising, then, that Squirrell, in the retrospect of seven years, should have written as follows:†

the small-pox, may we not infer that a mode of inoculation may be introduced preferable to that at present adopted, especially among those families which, from previous circumstances, we may judge to be predisposed to have the disease unfavourably?"

 * *A Treatise on the Lues Bovilla or Cow-Pox.* By Benjamin Moseley, M.D., Physician to the Royal Military Hospital at Chelsea. 2nd ed. London, 1805, p. 122.

 † *Loc. cit.*, p. 48.

" After meeting with these difficulties and ambiguities, and perceiving both the partial and general malignant effects of this virus on the milkers, with the uncertainty, at that time, of the change which, by inoculation, it would produce in the human system, I should have imagined that Dr. Jenner would have been totally prevented from commencing such an innovation. Perhaps, had he been better acquainted with the true mode of small-pox inoculation, prudence would have induced him to relinquish for ever a practice replete with such obscurities as render the practitioners liable to constant mistakes, and the public to reiterated disappointments ; at the same time promising no prospect of success or security. He could not be justified, either by experience or science, in pursuing such a measure. . . Had he but seriously reflected upon the specific nature of the cow-pox virus, and endeavoured to trace the grease of the horse, ' whence it sprang,' up to its true origin, previously to engaging in so momentous a concern, it would have required but a small share of abilities to have predicted the dreadful consequences that have since ensued ; and I am sorry to have the opportunity of observing that the result of the industry which he has shown, in ingrafting an unknown disease into the human constitution, neither merits private regard nor public approbation."

We shall see that Squirrell's accusation of ingrafting " an unknown disease " into the human constitution was a good deal nearer the mark than many easy-going and incurious persons have ever suspected.

CHAPTER IV.

WHAT IS COW-POX?

"If any man is ignorant of the origin or effect of the cow-pock,
is either owing to want of reading or of intellect."—JOHN RING,
Answer to Mr. Birch, in defence of Vaccination, 1806.

VACCINATION with lymph of the original stock was carried
on for many years in a hopeful spirit, although not without
some slackness from time to time and indifference on the
part of the public. It was not until between 1830 and 1840
that complaints began to be heard of the "degeneracy" of
the vaccine in those years, and proposals mooted for "going
back to the cow." One important result of that movement
was the inquiry by Ceely, of Aylesbury, on the natural
history of cow-pox, and by Estlin, of Bristol, on the effects
of "primary" lymph; while another and prior result was
the inquiry in Würtemberg, by a system of collective investigation stimulated by premiums, under the general
direction of Hering, the chief veterinary authority of that
State. I quote the following significant remark from the
preface of Hering's report: "We had been going on inoculating with cow-pox lymph, while we hardly knew what
the pox in the cow was like." This admission, which is not
less true of the generality of the profession now, than it
was then, or had been from the beginning, will enable us to
make the true application of Ring's two-edged dictum,
quoted at the head of this chapter, and to estimate the
fairness of his retort to Birch, when he accuses the latter (1)

of being ignorant of the origin of cow-pox; (2) of being ignorant of its effect; and (3) of being ignorant that small-pox is a scourge of the human race.

The real nature of cow-pox, or pap-pox of the cow, was well enough known to Jenner, although it is hard to believe that the same writer, who constantly admits its property of "corroding ulceration" or phagedena, should have invented for it the name of *variolæ vaccinæ*, or "small-pox of the cow." Writers subsequent to Jenner, always excepting Ceely, have given themselves very little concern with the actual facts and circumstances of the disease in the cow's teats and udder; they have accepted the name of "small-pox of the cow" at Jenner's hands implicitly, although a more monstrous perversion of sense or abuse of analogy is hardly to be found among the many instances of that vice in the history of medical doctrine. I am concerned here only with the natural history of cow-pox, and it is foreign to my purpose to enter upon the history and natural history of small-pox; but any one having the most modest acquaintance with epidemiology, not to speak of clinical medicine, will see on a moment's reflection that there is absolutely no parallelism between the general febrile eruption of the human skin, with remarkable contagious properties, known as variola, and the occasional outbreak of pimples, teased into ulcerations, on the teats of a milch cow or heifer here and there, under certain peculiar circumstances of season or of physiological constitution. We shall see in due course the full meaning of the deep scar of cow-pox: it means, on the face of it, a loss of substance through the whole thick-ness of the corium; and the pits of small-pox mean the same thing. But the circumstances of the loss of substance in the two cases are very different, just as the antecedents and natural history of the two diseases are very different. There is only one animal disease that runs on all fours with human small-pox, and that is the sheep-pox, or *variola ovina*. It is, like the former, a general pustular eruption of the skin ; it is contagious from animal to animal by effluvia ; it is often fatal, according to the amount of eruption ; it is a disease not only of ewes in milk, but of both sexes and of all sexual conditions; and if its epidemiological history and geographical distribution were inquired into, its diffusion would be found to have the same association with foreign breeds of sheep as human small-pox has had with tropical races of men.*

* See Simonds, *Practical Treatise on Variola Ovina, or Small-pox in Sheep* (London, 1848); and Fleming, "Human and Animal Variolæ," in *Lancet*, 1880, vols. i. and ii. Fleming's general point of view, as regards cow-pox, "horse-pox," sheep-pox, and human small-pox, will be sufficiently indicated by the following passage from his introductory paper (*loc. cit.*, i. 165) : "I believe every species has its own independent and particular kind of variola, and I am unable to understand why man and the sheep should alone have the unhappy privilege of being the subjects of different forms, and this privilege be denied to all other creatures." Accordingly, the historical facts of cow-pox, as I shall detail them in the sequel, are not inquired into by Fleming. Bollinger, also a veterinarian, and now professor of pathological anatomy in the University of Munich, is equally indifferent to the historical facts ; in a paper, "Ueber Menschen- und Thierpocken," in *Volkmann's Sammlung*, No. 116, 1877, he upholds the thesis that there is no such thing as original cow-pox, that the disease always rises by infection of the cow's teats from the hands of milkers, that in former times the source of such infection was small-pox, but at present is usually human vaccinia, which, as he says, is now universally distributed among mankind. The veterinary monograph by Fürstenberg, *Die Milchdrüsen der Kuh* (Leipzig, 1868), is not only excellent on the anatomy and mechanism of the cow's udder ; but, in the section on cow-pox, is also more realistic than the usual veterinary treatises, although it divides the disease, after the arbitrary fashion mentioned in the text

Jenner's bold transference of the term *variola* to the pap-pox of the milch cow was not merely a catachresis of speech, as Pearson remarked of it in his first essay ; * it was also a master-stroke of diplomacy. The notion that cow-pox is small-pox of the cow has taken a deep hold of the popular and professional mind, and has even warped the judgment of men whose observations in matters of fact were of the most accurate kind.

Ceely's own surrender in this field is the most significant of all. No one has so laboriously sought out and faithfully set down in plain words what the spontaneous and sporadic cow-pox really is, what the cow-pox communicated to other cows really is, and what the cow-pox communicated to the milkers' hands, arms, or faces really is. His two memoirs in the *Transactions of the Provincial Medical and Surgical Association*, vols. viii. and x., are among the most realistic of pathological recitals, all the more creditable to their author in that their fidelity to nature had to be maintained amidst the most discouraging surroundings, in the gloom and filth of cow-houses, at short or casual notice, while carrying on his practice, and at considerable intervals of time. Ceely's natural history of cow-pox in the Vale of Aylesbury is our most valuable repertory of facts, confirming in closest particulars whatever facts (as dis-tinguished from theories) Jenner has recorded of the natural history of cow-pox in the Vale of Berkeley.

at p. 65 and p. 82, into an essential papular and vesicular part, and an accidental phagedenic ulcerating part, due to the incessant irritation of the sore teats in milking. His account of cow-pox appears to he mostly a compilation, and by no means an exhaustive one.

* " Remarks on the term *Variolæ Vaccinæ*," in his *Inquiry concerning the History of Cow-pox.* London, 1798.

What, then, is meant in speaking of Ceely's "surrender" to the Jennerian fancy of *variolæ vaccinæ*? To make this clear enough for our present purpose we shall have to diverge for a few pages.

In the section of his first memoir entitled, "Affinity of Cow-pox and Small-pox," Mr. Ceely writes as follows: "Neither the protective nor the modifying power of the vaccine, its co-existence with variola in the same individual, the striking resemblance in the form and structure of the two vesicles at a given period, nor its occasional occurrence as a secondary eruption, had satisfied me of their common origin." All the points here enumerated, not even excepting the occasional vaccinal eruption, are, indeed, calculated to satisfy one of an entirely dissimilar origin. Woodville's early experience, on a considerable scale,[*] that the vaccine infection ran its course altogether unaffected by concurrent incubation of small-pox, and that the variolous infection, whether produced by aërial contagion or by inoculation, ran its course irrespective of concurrent vaccination, could only mean that the two infections were so far apart in their nature and significance as to want even that degree of reciprocal exclusiveness which some infections do show towards each other when they are implanted in the body together.

As for the "striking resemblance in the form and structure of the variolous and the vaccine vesicle," that was a piece of Jennerian slovenliness, or something worse, that ought to have been wholly got rid of long before Ceely's time. More than one early writer gave a picture of variola and vaccinia side by side in the several stages of their

---

[*] See chapter ii., *supra*, p. 26.

development, by means of which any one may see that the resemblance is only "at a given period," as Ceely says, namely, the stage of the papule, wherein most eruptions are on common ground, the subsequent development of each being widely divergent. I have already pointed out, in speaking of the inoculation-vesicle of horse-grease,* what significance the deep destruction of the corium has in the vaccine vesicle ; and I have suggested what it implies in the generalised eruptions of small-pox and sheep-pox, namely, an anatomical peculiarity in the skin of certain foreign races or breeds.†

Nothing could be clearer or more conclusive than Pearson's matter-of-fact exposure of the illusion about the vaccine vesicle being comparable to a pustule of small-pox ; ‡ but, like everything else that was reasonably urged in those days by way of criticism of the Jennerian doctrines, even if the criticism came from vaccinators themselves, it was set down to jealousy of the great man, and was met by the stolid opposition of the Jennerian interest. The only way to make out even a remote likeness between variola and cow-pox is to place a confluent patch of the former beside a rather irregularly shaped vesicle of the latter. Ceely, indeed, had the best of reasons for being sceptical as to the "common origin" of the pox of the cow's teats and udder, and the generalised skin disease of man known as variola.

However, in the course of a series of experiments to

---

* Chapter iii., *supra*, p. 34.   † *Supra*, p. 45.

‡ *Examination of the Report of the House of Commons Committee* (London, 1802), pp. 104—107. Among other systematic contrasts between the variolous pustule and the cow-pox vesicle, that of Auzias-Turenne (in the posthumous volume entitled *La Syphilisation*, Paris, 1879, p. 639) may be recommended as specially exhaustive.

retro-vaccinate the cow with direct cow-pox matter, or with humanised lymph, he came at length to try whether he could not variolate the cow, or inoculate it with small-pox matter. After several failures, he succeeded in raising a variolous pock, not on the skin, but on the mucous membrane of the vulva of one heifer, and four variolous pocks on the same mucous membrane of another. That the pock in the first was really variolous (it was of enormous size) was proved by an accident that happened with the lancet covered with matter from it still warm : the assistant, while holding the lancet, accidentally pricked his hand, and in due course developed the local variolous pustule, and the general variolous eruption with fever, notwithstanding that he had been vaccinated and had passed through a casual attack of small-pox as well. But, in some of the later removes from the parent pock, Ceely was able to raise vesicles on the child's arm, which were regarded as correct. He cultivated his lymph on the principle of judicious selection ; and the Vaccination Section of the Provincial Medical and Surgical Association reported as follows on the lymph from that source used at Cheltenham in 1840 : " The correctness of the vesicle formed by it exhibits a marked contrast to that which we have seen produced by other virus now in use." *

Attempts on the part of others to variolate the cow, and to raise vaccine thereby, have mostly failed : notably the attempt of the Lyons Commission, in 1863, directed by M. Chauveau,† and the attempt of Dr. E. Klein, in 1879,

---

* *Transactions*, viii. (1840), p. 26.

† *Vaccine et Variole.* Rapport par A. Chauveau, Viennois, et P. Meynet. Paris, 1865. A full analysis of this report will be found in

acting in concert with Mr. Ceely himself, at the instance of
the Local Government Board.* At the same time it is not
easy to detect any source of error in the section of Mr.
Ceely's first memoir, entitled *Variolation of the Cow:*
the pustules were raised, it should be remembered, on a
highly vascular semi-exposed mucous membrane, and one of
them spread out enormously to the breadth ; the matter
from one of the cases, accidentally inserted when still warm
on the lancet, produced a true variolous infection of the
human subject ; and, in the subsequent removes, the pock at
the place of insertion retained for several generations something of the dusky redness of the highly vascular soil on
which it had been raised in the heifer, as well as the
original look of purulence in the contents. It is quite
credible that a strain of lymph was cultivated, by judicious
selection, from that source, which produced only a single
vesicle or pustule at the point of insertion, and that vesicle
a large and umbilicated one. A success not unlike that was
attained, in the name of true variolation, by inoculators
such as Gatti,† and by Adams in his arm-to-arm inoculations
with the so-called pearly variety of small-pox.‡ That Ceely
should have succeeded in twice variolating the heifer
on a highly vascular mucous membrane, and of continuing a stock of matter therefrom, is quite intelligible ;
and it is equally intelligible that he and others should
have many times failed to do so both before and after that
success.

Fleming's series of papers on "Human and Animal Variolæ," *Lancet,*
1880.

    * *Rep. Med. Off. Loc. Gov. Board for* 1879, Appendix, p. 135.

    † See Bohn, *loc. cit.,* p. 76.   ‡ Baron, i. 246.

I am unable, therefore, to admit the suggestion that Ceely had mixed up his variolous and vaccinal lancets, and had really vaccinated when he believed he was variolating. The criticism should fall, not on the reality of his experimental procedure, but on the correctness of his theoretical conclusion. Having "managed" his variolous matter in such a way as to produce correct vesicles after a few removes, he concluded in defiance of common sense that the original cow-pox was "the vaccine modification of variola;" just as Jenner and others concluded, from the same evidence of correctness in the vesicle on the human arm, that the grease of the horse's hocks was the equine modification of variola. ·I have already pointed out, in the chapter on horse-grease as a source of vaccine, that the same sort of argumentation would show the venereal pox to be a modification of variola, or, in other words, that the great-pox and the small-pox are the same disease under different circumstances.

Still further, Ceely himself raised vesicles in man with one or more of the varieties of spurious cow-pox,† which could hardly be distinguished, so far as looks went, from the true Jennerian vesicle. Lastly, if the reader will take the trouble to turn to Hering's plate of the many forms of pox found on the cow's teats and udder, and if he will select the only one on the whole sheet which shows the much-talked-of depressed centre and the puffed or distended margin, he will find that the pock in question is the white or blister-pock, which all agree in regarding as spurious.‡

* Fleming, in a letter to the *Lancet*, 1886, ii. p. 999.
† See chapter vi., p. 94.
‡ *Ueber Kuhpocken an Kühen.* Stuttgart, 1839.

It is not the section on "Variolation of the Cow" that
constitutes the valuable part of Mr. Ceely's writings; it has
been unfortunate, indeed, for his scientific credit that he
was ever led into that side issue of experimentation. His
bizarre experiment was promptly seized upon as an easy
rationale of the empirical, not to say mysterious, efficacy of
vaccination; and there are probably few of the present
generation of medical men who know at all accurately what
Ceely's services to the study and elucidation of cow-pox really
were.   I feel bound to quote the following authoritative
reference to Ceely, which omits even to mention his laborious
studies of cow-pox itself.   In the well-known vaccination
blue-book of 1857 * we read as follows : "It was not until
forty years after that science supplied an authentic interpretation of Jenner's wonderful discovery.   .   .   .   These researches [by Ceely and others] set in a very clear light the
meaning of Jenner's practice.   A host of theoretical
objections to vaccination might have been met, or, indeed,
anticipated, if it could have been affirmed sixty years ago as
it can be affirmed now (1857) : This new process of preventing small-pox is really only carrying people through
small-pox in a modified form.   The vaccinated are safe
against small-pox because they, in fact, have had it."

It is true that all the exponents of vaccination have not
agreed in that doctrine, or they have not been steadfast to
it.   Thus Dr. G. Gregory, early in his career, was of opinion
that "vaccination is not small-pox, but just the reverse, the
antagonistic principle"; and Sir Thomas Watson came forward towards the end of his life to maintain that "the

* *Papers relating to the History and Practice of Vaccination.* Presented to both Houses of Parliament. London, 1857.

vaccine disease is *sui generis ;* the true attitude of cow-pox towards small-pox is an attitude of antagonism." The received doctrine of vaccination has, in fact, been blown about by every wind of fancy, and all for want of a little attention to the plain natural-history characters of cow-pox as they may be gathered from authentic sources, and from none more authentically than Mr. Ceely's own writings. We may now return to the inquiry from which we diverged, "What is cow-pox ?"

Jenner's account of the primary pox in the cow's teats is as follows :

"It appears on the nipples of the cows in the form of irregular pustules. At their first appearance they are commonly of a palish blue, or rather of a colour somewhat approaching to livid, and are surrounded by an inflammation. These pustules, unless a timely remedy be applied, frequently degenerate into phagedenic ulcers, which prove extremely troublesome."

He says no more than that in his first essay, with regard to the local characters of the pap-pox ; in his later writings the references to the cow's malady are merely incidental, and mostly for the purpose of supporting the distinction drawn by him between genuine and spurious cow-pox. One incidental remark in the second pamphlet is as follows :

"I have often stood among a herd which had the distemper without being conscious of its presence from any particular effluvia. Indeed, in this neighbourhood it commonly receives an early check from escharotic applications of the *cow leech.* It has been conceived to be contagious among cows without contact ; but this idea cannot be well founded, because the cattle in one meadow do not infect those in another (although there may be no other partition than a hedge) unless they be handled or milked by those who bring the infectious matter with them. .   . All attempts to communicate it by effluvia have hitherto proved ineffectual."

For the diagnosis of true from spurious cow-pox he gives the following criterion (which we shall see is directly contradicted by Ceely's wider experience) :

"These white blisters on the nipples they say (*i.e.* the dairy folks) *never eat into the fleshy parts* like those which are commonly of a bluish cast, and which constitute the *true cow-pox ;* but they affect the skin only, quickly end in scabs, and are not nearly so infectious."

Such is the sum of all the information ever published by Jenner concerning the pox as it occurs on the cow's teats. However, we have a good many more details from him concerning the same disease communicated to the milkers' hands, arms, or faces ; and that evidence does duty, in great part, for what is wanting concerning the parent disease. I transcribe the references, giving the general statements first, and the particular cases next :

Jenner on cow-pox as communicated to milkers.—" Inflamed spots now begin to appear on different parts of the hands of the domestics employed in milking, and sometimes on the wrists, which quickly run on to suppuration, first assuming the appearance of the small vesications produced by a burn. Most commonly they appear about the joints of the fingers, and at their extremities ; but whatever parts are affected, if the situation will admit, these superficial suppurations put on a circular form, with their edges more elevated than their centre, and of a colour distinctly approaching to blue. Absorption takes place, and tumours appear in each axilla. The system becomes affected, the pulse is quickened ; shiverings, succeeded by heat, general lassitude, and pains about the loins and limbs, with vomiting, come on. The head is painful, and the patient is now and then even affected with delirium.* These symptoms, varying in their degrees of violence, generally continue from one day to three or four, leaving ulcerated sores about the hands, which, from the sensibility of the parts, are very troublesome, and commonly heal slowly, frequently becoming phagedenic, like those from whence they sprung. The lips, nostrils, eyelids, and other parts of the body, are sometimes affected with sores; but these evidently arise from their being heedlessly rubbed or scratched with the patient's infected fingers."

I shall now collect, for the sake of the typical lesion, all the particular cases of accidental milkers' sores to be found in Jenner's writings:

(1) Joseph Merret, farm servant: several sores appeared on his hands soon after the cows became affected with the cow-pox; swellings and stiffness in each axilla followed, and he was so much indisposed for several days as to be incapable of pursuing his ordinary employment.

(2) Mrs. H——, a respectable gentlewoman: the infection was given to her through handling some of the utensils which were in use by those who had the disease from milking infected cows. Her hands had many of the cow-pox sores upon them, and they were communicated to her nose, which became inflamed and very much swollen.

(3) Sarah Wynne, dairymaid: she caught the complaint from the cows, and was affected with the symptoms (as above described in general) in so violent a degree that she was confined to her bed, and rendered incapable for several days of pursuing her ordinary vocations on the farm.

* The later editions have a foot-note here, to give expression to the sophistical after-thought of the second pamphlet (p. 103), distinguishing the "first action of the virus on the constitution," from that which "often comes on, if the pustule is left to chance, as a secondary disease." The note runs: "It will appear in the sequel that these symptoms arise principally from the irritation of the sores, and not from the primary action of the vaccine virus upon the constitution."

(4) Househo'd of Mr. Andrews, dairy farmer: all of them had sores upon their hands, and some degree of general indisposition, preceded by pains and tumours in the axillæ.

(5) Elizabeth Wynne, dairymaid: she had it in a very slight degree compared with her fellow-servants, one very small sore only breaking out on the little finger of her left hand, and scarcely any perceptible indisposition following it (seo No. 11).

(6) William Smith, farm servant; on one of his hands he had several ulcerated sores in 1780; in 1791 he caught the disease a second time; and a third time in 1794, equally severely as on the two former occasions.

(7) Simon Nichols, farm servant: hands began to be affected in the common way, and he was much indisposed with the usual symptoms.

(8) William Stinchcomb, farm servant: his left hand was severely affected with several corroding ulcers, and a tumour of considerable size appeared on the axilla of that side; right hand had only one small sore, no axillary swelling.

(9) Sarah Nelmes, dairymaid: a large pustulous sore on the hand (figured, at its vesicular stage, in Plate I. of the *Inquiry*), and the usual accompanying symptoms; also two small pustules on the wrists.

(10) Anonymous case "of a poor girl who produced an ulceration on her lip by frequently holding her finger to her mouth to cool the raging of a cow-pox sore by blowing upon it."

(11) Elizabeth Wynne (second attack in 1798): on the eighth day general lassitude, shiverings, alternating with heat, coldness of the extremities, and a quick, irregular pulse, pain in axilla; on her hand one large pustulous sore, in the vesicular stage, very painful.

(12) Young woman who had the cow-pox to a great extent, several sores which maturated having appeared on the hands and wrists; slept in the same bed with a fellow-dairymaid, but did not infect her.

(13) Instance of a young woman on whose hands were several large suppurations from the cow-pox; she at the same time nursed an infant, but did not infect the latter.

(14) Elizabeth Sarsenet, dairymaid : she had several sores upon her fingers, but no axillary swellings nor general indisposition.

(15) Hannah Pick, fellow-servant of the foregoing: she had sores upon her hands, and felt herself much indisposed for a day or two.

(16) Four or five farm servants, whom Jenner had failed to vaccinate by arm-puncture in the summer of 1798, all caught the disease within a month afterwards from milking the infected cows, and some of them had it severely.

Information collected by Pearson on the pox of the cow's teats and the milker's hands.—Almost immediately on the publication of Jenner's *Inquiry* in June, 1798, Dr. George Pearson, physician to St. George's Hospital, London, took up the idea with great ardour, and set about collecting information both on the natural history of cow-pox in the cow and in the milkers, and on its alleged protective power against small-pox in the latter. His queries addressed to correspondents were much more scientifically drawn and comprehensive in scope than those instituted by Jenner, although few of the answers were conceived in the same spirit. However, before the month of November, he had received a good deal of information from various parts of the country. The facts stated about the disease in the cow are so meagre, except as regards relative frequency, that they are hardly worth quoting ; I shall, therefore, proceed directly to his cases of milkers' sores :

(1) Thomas Edinburgh, cow-tender, Marylebone Fields, had cow-pox at the age of twenty ; was so lame from " the eruption " on the palm of the hand as to leave his employ, and become a patient in hospital for some time ; for three days had pain and swelling in the armpit ; disease was uncommonly painful and of long continuance. Has now (1798) a cicatrix in palm of hand, seen by Pearson.

(2) Thomas Grimshaw had cow-pox at the same time as No. 1; illness much less protracted: no other particulars.

(3) Information from a dairyman at a farm in the Hampstead Road: the milkers in Wilts and Gloucestershire sometimes so ill with cow-pox as to lie in bed for several days; a fever at the beginning; no one ever died of it.

(4) Information from a dairyman at farm near Somers Town, who "has a good understanding, is a man of veracity, and had lived in dairy farms all his life:" has seen the disease often in Somerset and in London; it affects the hands and arms of the milkers with painful sores, as large as a sixpence, which last for a month or more, so that they have to give up work.

(5) Information from Mr. Dolling, of Blandford: swelling under the arm, chilly fits, and other early symptoms; after the usual time of sickening, namely, two or three days, there is a large ulcer, not unlike a carbuncle, which discharges matter. (Dr. Pulteney, of Blandford, wrote in almost identical terms.)

(6) Information from Mr. Rolph and Mr. Grove, of Thornbury, Gloucestershire: hundreds of instances of cow-pox in milkers, but never a mortal or even dangerous case; the patients were ordinarily ill of a slight fever for two or three days, and the local affection was so slight that the assistance of medical practitioners was rarely required.

(7) Evidence of Mr. Fewster, also of Thornbury, Gloucestershire: in the course of thirty years had known numberless instances of the disease, but never knew one mortal, or even dangerous case; but thinks it a much more severe disease in general than the inoculated small-pox.

(8) Evidence of Mr. G. G. Bird, relating to the neighbourhood of Gloucester: milkers' cow-pox appears with red spots on the hands, which enlarge, become roundish, and suppurate; tumours arise in the armpits, the pulse grows quick, the head aches, pains are felt in the back and limbs, with sometimes vomiting and delirium.

(9) Evidence of Mr. Wales, Downham, Norfolk. (*a*) Case of old farm servant, who had cow-pox in early life: "eruption" on his hands considerable, and his fingers swollen; the places healed slowly, and

left scars which are evident at this day; and when the hands are very cold these scars are of a livid cast.

(b) Has met with two cases in which the matter of the cow-pox, by being applied to the eyes, destroyed the power of vision from the opacity of the cornea so produced.

(c) No person has been known to die, or even to be in danger with the cow-pox, although the axillary glands have been much affected, and the sores on the hands have healed with difficulty.

(10) Evidence of Dr. Fowler, of Sarum: "This morning (24th of October, 1798) Anne Francis, a servant-girl, aged twenty-six, was brought to me; she informs me that some years ago bluish pustules arose on her hands from milking cows diseased by the cow-pox. These pustules soon became scabs, which, falling off, discovered ulcerating and very painful [sores], which were treated by a cow doctor, and were long in healing."

From the evidence sent in to him, Pearson concluded that the milkers' cow-pox was of two degrees: (1) those cases where they are confined to bed for several days, and have "painful phagedenic sores" for several months; and (2) cases so slight that the patients are not confined at all, but get well in a week or ten days.

Ceely's original observations of cow-pox on the cow's teats and on the milker's hands.—Besides the information collected by Jenner and by Pearson, hardly anything was added to the natural history of cow-pox in England until Estlin and Ceely wrote upon the subject from 1838 to 1842. The English inquiries in the years immediately following Jenner's announcement, as well as most of the foreign ones, were devoted mainly to discovering a basis in experiment for the horse-grease hypothesis. In the vaccination controversy nothing new was adduced as to the characters of cow-pox in the cow, or in the casually communicated human form; the

negative writers, such as Moseley and Squirrell, resting
their case on Jenner's own data, while the apologists, such
as Ring, employed their whole skill in evading the questions
naturally suggested thereby.   The profession at large were
supplied with lymph which, as at present, is only occasionally
a source of serious mishap, and they were profoundly indifferent whether ulceration, and secondary infection, were
part of the complete natural history of cow-pox or not.

Ceely takes the facts of cow-pox in what may be called
their physiological order : (1) the spontaneous and sporadic
disease in the cow ; (2) the disease as communicated to the
teats of other cows in the same shed by the medium of the
milkers ; and (3) the disease on the milkers' hands or faces.

The spontaneous or sporadic disease was hardly ever seen by
Ceely himself at or near the commencement ; he had to depend on what
he was told for his knowledge of its circumstances, and of its first
symptoms.   When an outbreak occurs in a cow-house, the milkers
pretend, in general, to point out the infecting animal ; many intelligent
dairymen believe that it occurs more frequently as a primary disease
among milch heifers.   The following are two cases : In December,
1838, in a large dairy, a milch cow slipt her calf, had heat and induration of the udder and teats, with vaccine eruption, and subsequent
leucorrhœa and greatly impaired health ; the whole of the dairy, con-
sisting of forty cows, became subsequently affected, as well as some of
the milkers.   In another dairy it first appeared in a heifer, soon after
her calf was weaned ; in about ten or twelve days it was communicated
to five other heifers and one cow in the same shed, the milkers being
also affected.

Of the spontaneous or autochthonous development of
this malady in a cow now and again, Ceely makes no question.   Several times, in his experience, the disease, when it
had been once started in that autochthonous manner, went

through a whole cow-house; but such accidents were separated by intervals of years, and would occur now at one farm, now at another. The true cow-pox of Ceely's description is, indeed, the spontaneous cow-pox whose existence, at least, was recognised not only by the dairy folks, but also by medical writers, including Jenner himself. Jenner's hypothesis of the horse-grease origin of cow-pox led him to place the spontaneous cow-pox among the spurious forms of that disease; but he understands by spontaneous cow-pox the same malady which Ceely has fully described under that name. The following is Jenner's reference to it :

" But first it is necessary to observe that pustulous sores frequently appear spontaneously on the nipples of the cows, and instances have occurred, though very rarely, of the hands of the servants employed in milking being affected with sores in consequence, and even of their feeling an indisposition from absorption. These pustules are of a much milder nature than those which arise from that contagion [*i.e.* horse-grease], which constitutes the true cow-pox. They are always free from the bluish or livid tint so conspicuous in the pustules of that disease. No erysipelas attends them, nor do they show any phage-d·nic disposition as in the other case, but quickly terminate in a scab, without creating any apparent disorder in the cow. This complaint appears at various seasons of the year, but most commonly in the spring, when the cows are first taken from their winter food and fed with grass. It is very apt to appear also when they are suckling their young."

Jenner's other reference to spontaneous cow-pox is in *Further Observations* (*ed. cit.*, p. 77) :

" That which appeared to me as one cause of spurious eruptions, I have already remarked in the former treatise, namely, the transition that the cow makes in the spring from a poor to a nutritious diet, and from the udder's becoming at this time more vascular than usual for the supply of milk." He then mentions another cause, the interruption of milking when a cow is exposed for sale: "thus the milk is preternaturally accumulated, and the udder and nipples become greatly distended. The consequences frequently are inflammation and eruptions which maturate. Whether a disease generated in this way has the power of affecting the constitution in any *peculiar* manner I cannot presume positively to determine. It has been conjectured to have been a cause of the true cow-pox, though my inquiries have not led me to adopt this supposition in any one instance; on the contrary, I have known the milkers affected by it, but always found that an affection thus induced left the system as susceptible of the small-pox as before."

Although Ceely, as we shall see (p. 69), was not free from the same *arrière pensée* that Jenner very plainly betrays in the last sentence of the quotation, in distinguishing the true cow-pox from several varieties of " spurious ; " yet his true cow-pox was the spontaneous and sporadic, and he knew of none in his district which had an origin in the grease of the horse. *

So far as the facts could be learned about the primarily affected cow or heifer, the animal had heat and tenderness of the teats and udder for three or four days ; then followed a pimply hardness of the parts, the pimples being of a red colour, quite hard, and as large as a vetch or pea ; in three or four days many of them will have grown to the size of a horse-bean ; " the tumours rapidly increase in size and tenderness, and some appear to run into vesications on the teats, and are soon broken by the milker's hands." Crusts form on the sores, which get detached at short intervals by the "merciless manipulations of the milkers; " and an inveterate state of soreness arises.

It is impossible to follow the fortunes of the "natural" disease farther, as distinguished from the "casual," or that which is transmitted from cow to cow by the milker's hands ; but as the latter would seem to reproduce the type of the

* *Trans. Prov. Med. and Surg. Assocn.* viii. (1840), pp. 299, 300.

primary affection, just as the sores on the milker's hands do, it is of less consequence to consider the pathology of each apart.

The spectacle that usually met Mr. Ceely's eye was that of a cow-house full of animals in all stages of the disease at once; such was the Dorton case, graphically described in his second memoir.

Around the bases, or on the bodies of the teats, the cows that had been first affected had scars of various sizes up to that of a chestnut, some of them puckered and uneven, and all of them with their outer edges slightly elevated and gently rounded off, and the bases a little indurated. In other animals the cicatrisation was incomplete; the rounded margin and the induration of the base were more conspicuous, and the centre contained small florid granulations or crusts. In still other cows, not so far through the process, there were ulcers of all sizes in a granulating state, sometimes covered with a crust, and always with hard bases and rounded margins. Crusts of all shapes and sizes entered prominently into the picture; and, lastly, there were the vesicles. With reference to the vesicles, it would be a mistake to suppose that they resembled the vaccine vesicle of the human skin; they were, for the most part (and this is especially true of the genuine as distinguished from some varieties of the spurious), small collections of fluid under a thick skin at the very apex of a large and hard pimple. Indeed, perfect vesicles, or the disease in its vesicular stage, would appear to be a mere ideal; "generally the majority of the tumours are more or less abraded or otherwise injured, either by the animal while recumbent, or by the merciless manipulations of the milkers." It is significant also that, for the practical purpose of inoculation, Ceely had to be content with crusts, owing to the difficulty of getting fluid lymph from the "eruption" on the udder.

The contrast with human vaccine is thus stated by Ceely himself:

"In man the containing cells, readily distending, elevate the yielding and thinner cuticle, whereas in the cow the lymph is slowly and scantily secreted for a time, the cuticle is thick and resisting, and an epidermic fissure affords the readiest outlet. A near approach to this tumour-form sometimes, it is true, is found in children in particular states of health, or in those of phlegmatic habits, otherwise healthy, with thick skins, where the vesicle, of a rose or damask hue, rises boldly, and in a solid form, above the level of the skin, covered with an ash-coloured or bluish epidermis, which being punctured, like that on the cow, yields scarcely anything but blood even till the tenth day."

Among the miscellaneous points noted in the natural history by Ceely are the following :

" In passing through a large number of cows it has appeared to me generally milder in the latest than in the first subjects, and I have certainly succeeded in effecting a mitigation by artificial means while in the prosecution of experiments with another view.

" Its topical severity depends almost wholly on the rude traction of the milkers.

" There is no derangement of health either in the animal primarily affected or in those secondarily affected. . . . The animal continues to feed and graze apparently as well as before.

" In the same dairy at the same time with the true disease, some one or other of the spurious forms may occur in some individuals [cows], causing difficulty in milking, and producing deep sores on the milkers' fingers, thus complicating the investigation and deceiving the indiscriminating milkers.

" Occasionally warty or fungous growths succeed some of the deeper ulcerations."

Ceely's description of the pox of the milch cow's teats is of the realistic order. It is commendably free from theoretic bias, and it is by far the richest in detail of any that has been given ; moreover, it bears out the original data of Jenner, and the average testimony of the country, as collected by Pearson. We may sum it up briefly as follows : An occasional breaking-out on the skin of the teats and udder, associated with the exercise of function by that great organ of secretion, and particularly with the changes incidental to it. The eruption is of hard pimples, mostly of large size, which "maturate" to a very limited extent where they pursue their natural development. Their "natural" development, in so far as it can come into the pathology of infective disease, does not exist, for the reason that the pimples on the cow's teats, if they were saved from the "merciless manipulations of the milkers," would simply run the course of pimples, and would never become pox ; it is the perpetual "insult" of an ailing part, the forcible traction on the pimply skin three times a day, the creation of hæmorrhagic crusts, and the ever-renewed displacement of these, that now and then sets up the inveterate and communicable process which we know as cow-pox.

Its characters are deep or spreading ulceration (sometimes phagedenic to a degree that destroys half the udder),* with slow healing, induration of the base and roundness of the edges, and a deep permanent scar, often smooth and regular, but not rarely puckered and irregular, such as follows any ulcerative destruction through the whole thickness of a vascular and almost erectile skin.

On the basis of these facts (for we are all alike dependent on Ceely's information) various eclectic accounts of cow-pox have been drawn up to suit the doctrine of *variolæ vaccinæ*, or small-pox of the cow. Thus Seaton, while professing the most unbounded confidence in Ceely's accuracy, and, indeed, reproducing his details with commendable fidelity, endeavours to separate "the uncomplicated disease" from

---

* Furstenberg, *Die Milchdrüsen der Kuh.* Leipzig, 1868.

E

" the local affection as disturbed by handling." But should we ever have heard of the affection at all, if it had not been disturbed by handling? and would it ever have become communicable from the sporadic primary cow to milkers, or transmitted to other cows, but for the same reason? Cow-pox is cow-pox just because the local affection *has* been disturbed by handling; the sores had been provoked, and made indurated or phagedenic, or otherwise inveterate, by nothing else than the merciless routine of milking, which did not allow them to heal; and the inveteracy in the individual has its complement in the communicability to others, or in the acquired specificity. Therefore, if we are to follow Seaton in defining cow-pox as a " specific disease," it is not the eruptive part of it that is specific: for that is merely the local expression of a constitutional derangement incident to the season and to the periodic function. The specific element comes in with the inveteracy; and, like specificity in some other diseases, it owes its existence to provocation, or to neglect and indifference to the reparative process, carried beyond the safe limit.

That we have no warrant to detach the ulcerative or phagedenic part of cow-pox from its pimply or slightly vesicular phase, and to make the latter the essential or uncomplicated disease, will be seen at a glance when we consider the disease as it is accidentally communicated to the hands or the face of a milker. There, at least, the incessant traction on the teats every six hours, with disturbance of crusts and provocation of sores, does not come in. And yet the disease goes through all the phases: it is a vesicle before it is an ulcer, just as an inoculated venereal sore is a vesicle before it is an open sore; but so far as we know the history of these milkers' sores, they never stop at the vesicular stage, or get healed without passing through the phase of an ulcer, which is usually painful and phagedenic at first, and perhaps indurated afterwards.

Jenner's cases of milkers' sores have already been transcribed in full; they included one case of a girl who got what was practically a cow-pox chancre on her lip from blowing upon her sore finger.   I shall now give Ceely's :

Ceely's cases of accidental cow-pox in the milkers.—1. Case of Mr. Pollard, aged fifty-six, 23rd Oct., 1840 : when first seen by the surgeon the vesicles on the hands and fingers had burst.   The patient stated that, about ten days after the discovery of the disease on the cows, he observed two itching small pimples on the site of the present ulcers; subsequent pain and tenderness of the axillary glands, with the usual constitutional symptoms, which increased for four or five days, but were never severe enough to confine him to the house; the sores, when seen by Ceely, had blackish-brown sloughs in their centres, and their bases were surrounded with an elevated induration of a livid red colour.

2. Joseph Brooks, aged 17, noticed the disease in three places almost at once—on a finger, on the thumb, and on the right temple, with premonitory tenderness of the lymphatic glands in the neck. First noticed as red pimples on 19th Oct., 1840; on the 21st he had headache, malaise, and axillary pain and tenderness, symptoms which increased during the next two days, when there was also nausea with vomiting; at the same date his right eye was closed by the surrounding swelling.   Coloured plates are given of the three vesicles as they showed on the 23rd Oct.: that on the temple is a large oval vesicle, half an inch long, fallen in at its centre, which bore a small brownish crust; the vesicle is tense and glistening, of a rose-red colour, and with a vivid areola.   The vesicle on the back of the thumb is about the same size, of a lemon-yellow tint, and seated on a raised base; that on the radial side of the ring finger is smaller and more pearly, but with a deeper zone of red round it, and an elevated base.

Lymph was got in small quantity, and with difficulty, from the vesicle on the temple, which was solid and compact like the initial papule of the teat; the vesicles on the finger and thumb were left unpricked, so that they might pursue their natural course. The development had proceeded considerably on the 26th and 27th Oct., the redness and swelling around had declined, and the vesicles become greatly enlarged; those on the thumb and finger were " loosely spread out at the circumference," each having a dark central slough; that on the temple was nearly an inch in diameter, it had a firm and fleshy margin, as in the cow, and a dark brown slough firmly adherent in the centre. After seven or eight days' poulticing, the sloughs separated, and the deep ulcers healed, leaving cicatrices which, on the 27th Nov. and 5th Dec., were found to be deep, puckered, and uneven.

3. Joseph White, aged 18, farm servant at Dorton, noticed pimples on thumb and back of the left hand on 25th May; constitutional disturbance and axillary tenderness for several days from 30th May to 5th June. Vesicles seen by Ceely on 2nd June and following days, and coloured drawings made; they went through the same stages as in case 2, the patient being kept at Aylesbury under Mr. Ceely's treatment; they became greatly enlarged, threw off central sloughs, and granulated slowly; on 12th June (the nineteenth day from first sign of pimples) "the stage of ulceration was fully developed, and the extent of topical disorganisation was now sufficiently manifest." After about a fortnight the ulcers had healed, "leaving scars like those succeeding variolæ, *or any other disease attended with entire destruction of the corium."* Plates are given of the three stages (papules on a swollen base, large vesicles, and ulcers), the ulcer on the thumb being upwards of half an inch in the long axis, level at the edges, deep in the centre, and covered all over with bright red granulations, while that on the back of the hand has a rounded raised margin of induration, is deeply excavated, and shows a bacon-like smooth floor.

Besides these details of cases, Ceely gives some general characteristics of the sores on the milker's hands or face.

He begins by saying that the dairy people often mistake

the spurious for the true cow-pox, on the hands or elsewhere, . making their diagnosis " mainly on the grounds of severity and communicability." Such spurious sores on the hands of milkers, " producing considerable local irritation, and much constitutional disturbance, often interrupting their avocations, and occasionally confining them to bed, have not only proved the source of much misinformation, but have imparted to many a confident assurance of safety from smallpox, which subsequent attacks of that disease have proved entirely unfounded." *

What follows relates to sores, adjudged true according to Mr. Ceely's criteria. They occur mostly on the backs of the hands, particularly between the thumb and forefinger, about the flexures of the joints, and on the palmar, dorsal, and lateral aspects of the fingers. " The forehead, eyebrows, nose, lips, ears, and beard are often implicated from incautious rubbing with the hands during or soon after milking." Women may have the sores on the bare part of the forearm. The central depression of the vesicle is not constant ; nor is the bluish colour, which evidently depends upon and is influenced by the vascularity of the part, the greater or less translucency of the epidermis, the quantity of lymph, the depth and extent of the vesicle. Although

---

* Ceely defines true vaccinal ulcers as follows (*loc. cit.*, x. 229) : " Vaccine ulcers are generally distinguishable by a rounded elevation, more or less manifest, of their outer margin, and a circumscribed induration of greater or less extent of their base, with a proportionate depression in their centres of deeper ulcerations, sometimes caused by a slough. . . If we can be positively assured that the above-mentioned diagnostic conditions have existed in any given ulcer for three or four weeks, or even longer, especially if it be removed from *severe* casualties, we may fairly presume that it is vaccine."

Cecly records no case where the disease in the milker healed
by shrivelling and drying of the vesicle, or under a scab,
and although Jenner specially claims the latter mode as
characteristic of spurious varieties, yet it is right to say
that the former observer does not speak of ulceration and
phagedena as the invariable rule. "The vesicles are *frequently* broken, or, when the epidermis is thin, spontaneously burst, causing deep sloughing of the skin and cellular tissue, and ulcerations which slowly heal. There is
often, consequently, much attendant local irritation, and
considerable symptomatic fever." Papular, vesicular, and
bullous eruptions are occasionally seen attendant on casual
cow-pox, especially in young persons of sanguine temperament and florid complexion, at the height or after the decline of the disease.

Such, then, is the real nature of cow-pox in the cow,
and in the first remove from the cow, when accidentally
inoculated on the milker's hand or face. We come next to
consider whether the mimicry of infection pursues the cow-
pox in its experimental production on the human arm ; or to
what extent, and under what circumstances, the traditions
of its native soil are forgotten or mitigated in the course of
humanising.

CHAPTER V.

THE EFFECTS OF VACCINE INOCULATION IN THE FIRST REMOVES FROM THE COW.

HAVING tried, in the foregoing chapter, to see cow-pox as it really is, both on the cow's teats and in its accidental form on the milkers' hands or faces, we may next seek, with no small curiosity, to discover whether the infection retains those remarkable characters, or how much of them it loses, in the experimental or systematic practice of vaccination.

In disinterring the original Jennerian experiments in the first chapter, I have already had occasion to notice the actual consequences of using primary lymph, no matter how the experimenter explained them away afterwards with his opportunist doctrine of primary and secondary effects. So much did he at an early period look upon ulceration as an integral part of the natural history (knowing it to be so in the cow and in the milker), that Cline, to whom he gave the first vaccine matter for trial in London, writes to him of "the ulcer," as if it were a matter of course and a termination that he had expected. Jenner, indeed, suppressed the reference, and in other ways tried to extenuate that awkward part of his early experience. But the very next attempt that he himself made, on December 2nd, 1798, brought him face to face with the difficulty in a worse form than ever, the vesicles in the two vaccinated children turning to phagedenic sores, which spread to the size of a shilling, and healed slowly.

He was still revolving in his mind that fatal barrier to the realisation of his dreams, when he was surprised by Woodville's London successes. He made trial of Woodville's lymph, and wrote of it at once : " The character of the arm is just that of cow-pox, except that I do not see the disposition in the pustules to ulcerate, as in some of the former cases."

Before that way out of his difficulty had been found for him in an unexpected quarter, he had resolved to boldly face and anticipate the objections to cow-poxing on the score of its ulcerous tendencies. Writing to Woodville in the end of January, 1799, he says : " I am shortly going to publish an appendix to my late pamphlet, to mention the precaution of destroying the pustule," etc. This letter was answered by Pearson, at Woodville's request : " On telling Dr. Woodville that I had been anxious about your publishing the use of the caustic, he replied, ' That would have damned the whole business.' Be assured that if the practice cannot be introduced without the caustic, or call it by any other name, it will never succeed with the public." *

Jenner, however, carried out his intention ; and we find in the *Further Observations* the promised directions about " destroying the pustule " by caustic, as well as certain theorisings on the liability of vesicles to " degenerate." He boldly published his two last cases of phagedenic ulceration, making them the text of his remarks. Woodville, by a stroke of fortune unequalled either before or since, had overcome the initial difficulties of vaccinating direct from the cow, and had provided Jenner and the rest of the world with a lymph which promised

* Baron, i. 315.

to have no bad consequences. Under cover of that safe practice, Jenner saw his opportunity to promulgate the doctrine of the spurious vaccine vesicle, which has played so great a part in the history of vaccination and of its mis-adventures. This will be the most convenient place to say something of that doctrine.

The doctrine of spurious or degenerate vaccine vesicles.—The doctrine of the spurious vaccine vesicle should not be confounded with the parallel doctrine of spurious cow-pox in the cow. Jenner's original spurious category was a very simple, if a very arbitrary one; it included every form of sore or vesicle on the cow's teats that did not arise by mediate contagion from the horse's hocks.* But in the second pamphlet " the sources of a spurious cow-pox " take a wider range, becoming four in number. The first is pustules on the cow's udder "which contain no specific virus"—a conveniently vague class; the second is matter which had originally possessed the specific virus, but had suffered decomposition either from putrefaction " or from some other cause less obvious to the senses ; " the third is "matter taken from an ulcer in an advanced stage, which ulcer had arisen from a true cow-pock ; " and the fourth is matter produced on the human skin from some peculiar morbid matter generated by a horse (*i.e.* not the real horse-grease). These were all " spurious," among other reasons because they corresponded to certain instances in the experience of Jenner and of others, where the promised protection against small-pox had not been made good.† It

---

* *Inquiry*, etc., 1798, p. 7.

† Jenner's original limitation of "genuine " cow-pox to an affection which came from the greasy hocks of horses, had the same motive behind

is only the third, the "matter taken from a true cow-pock ulcer *in an advanced stage*," that we have to do with here.

Enlarging upon his third head in the subsequent text Jenner says : "When this pustule [the true cow-pox pustule] has degenerated into an ulcer (to which state it is sometimes disposed to pass unless timely checked), I suspect that matter possessing very different properties may sooner or later be produced ; . . . and thus, by assuming some of its strongest characters, *it would imitate the genuine cow-pox*" (italics mine). Such, indeed, had been his own frequent, if not constant, trouble, until he obtained stock from Woodville's more benign lymph ; and such was the occasion of his introducing the precaution about caustics, much against Woodville's wishes. He then gives the case of Susan Phipps, whose vesicle became a phagedenic ulcer the size of a shilling, and that of Mary Hearn, in whom also "the progress of ulceration" had to be checked by mercurial ointment. He proceeds to say that, conceiving these cases to be important, he had given them in detail, firstly, to urge the precaution of using such means as may stop the progress of the "pustule ;" and, secondly, to point

it, namely, the familiar experience of his medical colleagues in Gloucester-shire, that cow-pox, as ordinarily defined and understood, had often failed to protect the milkers from small-pox. The following is from Baron's *Life*, i. 48 : "Dr. Jenner has frequently told me that at the meetings of this Society [the Convivio-Medical], he was accustomed to bring forward the reported prophylactic virtues of cow-pox, and earnestly to recommend his medical friends to prosecute the inquiry. All his efforts were, however, ineffectual ; his brethren were acquainted with the rumour, but . . . most of them had met with cases in which those who were supposed to have had cow-pox had subsequently been affected with small-pox." Jenner accordingly made the cases of "genuine" cow-pox a narrow class, to suit these numerous exceptions.

out that the vesicular process is all that need arise primarily, or from the "first" action of the virus, and that the
ulcerative or phagedenic process comes on "as a secondary
disease, if the pustule is left to chance." His own cases
with Woodville's lymph had doubtless helped him to that
important distinction, ever after observed in the theory of
vaccination. "As the cases of inoculation multiply," he
writes,* a week or two after his first successful series of
cases (with Woodville's lymph), "I am more and more convinced of the extreme mildness of the symptoms arising
merely from the primary action of the virus on the constitution."

Whenever we come to the really critical point of this
argument, Jenner goes off to the analogous case of bad
effects following small-pox inoculation, which bad effects,
according to Pearson, had no existence in the sense that
Jenner meant.† Disregarding this application, however,
we may take his admissions about cow-pox as significant in
themselves: "The simple virus [of cow-pox] itself, when it
has not passed the boundary of a vesicle, excites in the
system little commotion. Is it not probable the trifling
illness thus induced may be lost in that which so quickly,
and oftentimes so severely, follows in the *casual cow-pox*
from the presence of corroding ulcers?" These several
degrees of cow-pox virulence are naïvely adverted to as if
for the purpose of throwing light upon the supposed fact
(but actual error) that persons who had suffered from
small-pox cannot be vaccinated. The passage is interesting,

* *Further Observations* (April, 1799), *ed. cit.*, p. 109.

† *Examination of the House of Commons Committee's Report* (London,
1802), pp. 94, 95.

however, as an oblique admission that such degrees of virulence in the effects of cow-pox matter did really exist.

The " degeneration " of the vaccine vesicle into an ulcer was so notorious in Jenner's original experiments that Woodville and Pearson have each pointedly remarked upon the absence of that development in their own practice on the large scale. Woodville's statement at the end of his *Reports of a series of Inoculations for the Cow-pox* (1799) has been quoted before, but it has historical interest enough to be repeated : " We have been told that the cow-pox tumour has frequently produced erysipelatous inflammation and phagedenic ulceration ; but the inoculated part has not ulcerated in any of the cases that have been under my care. . . . It would seem, then, that the advantages to be derived from substituting the cow-pox for the small-pox must be directly in proportion to the greater mildness of the former than the latter disease."

Pearson has the following : * " Another correspondent (*Med. and Phys. Journal*, iv. 326), on the authority of Dr. Jenner, replies that it is fully ascertained that at a certain undetermined period, but always a late one, the cow-pock ' virus ' is capable of producing morbid and phagedenic ulceration, considerable erysipelatous inflammation, and a train of effects wholly dissimilar to those of pure and recently-formed virus." He gives his own experience as follows : † " As to phagedenic ulcers, as they have been called, ensuing from the inoculated part, many sore arms have been produced ; but nine out of ten were occasioned, or at least much aggravated, by the tightness of the clothes, by allowing the linen to stick to the sore, by scratching the pustule, and sometimes by

* *Examination of the Report*, etc., p. 121.   † *Ibid.*, p. 55.

emollient poultices. The experience we have had, then, since January last (1802), in London and in the country, does not exactly agree with Dr. Jenner's account concerning the state of the arms ; he thinks some new applications of a caustic nature necessary, in many cases, to prevent secondary symptoms from the sores ; but in Dr Woodville's *Report* (p. 155), my correspondents ' and my own practice, there has not been found any want of applications for such a purpose."

My only other reference to the point shall be taken from the *précis* of evidence given before Admiral Berkeley's Parliamentary Committee, in 1802 : Dr. Rowley, in answer to questions, "has not seen many cases of spurious cow-pox ; he has seen ulcers succeed in the beginning of vaccine inoculation, but that has been entirely obviated by the subsequent practice ; he does not know by what change in the practice these disastrous circumstances are now pre-vented."

Rowley afterwards explained * that he was made, in the *précis* of evidence, to say what he did not believe, adding that "ulcers, very bad ulcers, appeared afterwards, to which I have been witness, and have cured the cases by bark and vitriolic acid." At all events, to use his own language, "objectionable circumstances of a disagreeable nature were peremptorily said by ingenious vaccinators to have been removed ;" so that we may take their existence in the early practice as formally admitted.

The term "spurious," applied to any vaccinal process that went beyond the stage of a vesicle and scab, continued to be used according to the sophistical doctrine laid down

* *Cow-pox Inoculation no Security*, etc., London, 1805, p. 9.

by Jenner. It was not until the more candid researches into the effects of primary lymph by Ceely and others thirty or forty years later, that a scientific reason could be given for the fact that the vaccine process in the child's arm sometimes "imitates," as Jenner says,* the original corroding process of the cow's teats. I shall take Ceely's observations first, although they were preceded in time by those of Bousquet and Estlin :

Ceely's vaccinations with primary cow-pox matter.—Enumerating the distinctive character of vaccine vesicles produced direct from the cow, Ceely says : " The process of shrivelling, even in perfectly normal vesicles, is generally protracted. Although not so late in the thick clear skins of infants and some young children, even in these the drying-up process will be seen for some days to be confined to the centre, while the circular margin remains of a dull or dirty yellowish-white or pale horn colour, retaining a fluid *to the sixteenth or eighteenth day.* When the regular vesicle is neither ruptured nor spontaneously bursts, the crust is often retained to the end of the fourth or fifth week, bringing away with it a circle of the corium, often the whole depth of it, and some of the subjacent cellular tissue, leaving a deep foveolated red cicatrix, or a yellow foul excavation which ultimately furnishes the pink, shining, puckered scar. But it too often happens, especially in subjects with thin and vascular skins, that the vesicles burst, or are easily broken, during the height or about the decline of the areola ; and if the subject be of a strumous or erysipelatous diathesis of full habit, and possess an irritable skin, secondary inflammation is set up, and becomes more diffused and deeper seated : the corium is destroyed completely, and a slough of the subjacent tissue is soon manifest, the surrounding integuments are deeply indurated, often a multitude of echthymatous pustules are formed on the enlarged papillæ, and on other parts of the skin, and abscesses in the cellular membrane and axillary glands ensue, causing proportionate

* " And thus by assuming some of its strongest characters it would imitate the genuine cow-pox."—*Further Observations, ed. cit.*, p. 89.

constitutional irritation. When the slough separates, the wound often has the appearance of a caustic issue, seeming capable of receiving a small marble. All this mischief, however, generally soon subsides; the ulcers speedily clean, throw up luxuriant granulations, needing repression; the surrounding irregularly and superficially denuded skin soon heals, and an unexpectedly small circular or oval, red, shining, puckered, elevated, and uneven cicatrix succeeds."

The vesicles of primary lymph, although not unfrequently they are less fine and much less developed than other vesicles, "admit of very remarkable improvement by transmission of the lymph through a series of well-selected subjects. By this process, also, in a very short time, most of the defects and some of the evils connected with the use of primary lymph may be dissipated, and the lymph rendered milder and more suited to general purposes. . . By a steady and judicious selection . . . in a few (even three or four) removes, the severity of the local mischief becomes manifestly materially diminished, . . . and the lymph may be transferred with safety to others even more sanguine and robust," *i.e.* more so than the smooth and clear-skinned dark infants chosen to start the series with. But objectionable subjects have always to be carefully prepared, just as subjects used to be prepared for variolation; and some must even be refused altogether.

"In the succeeding removes, among a diversity of subjects, there is, of course, endless variety in the character of the vesicles. . Every now and then we have all the characters of the earlier removes, and all the inconveniences of primary lymph." Although the greater part of his experiments with primary lymph, and with lymph in the earlier removes, have exhibited the above as its qualities and accidents, he thinks it not improbable that primary lymph itself may not always be of the same strength as it comes from the cow; and he has observed the interesting fact that in passing through a large number of cows it has appeared generally milder in the latest than in the first subjects.

Such being Mr. Ceely's experience with primary lymph (including "a yellow foul excavation," even when the vesicle did not get broken, or did not burst spontaneously), it is not surprising that he was no great advocate for "going back to the cow." "My own repeated applications to the cow," he says (*loc. cit.*, viii. 376), "have been chiefly for the purpose of experiment, for the satisfaction of patients, or the accommodation of friends, not from any belief in its superior protective efficacy over active humanised lymph."

Jenner also discouraged attempts to go back to the cow, cordially accepting the lymph of Woodville as a peculiarly happy way out of the dangers that seemed at one time to beset the practice. Birch,* the opponent of vaccination, asks (1807) "Why are we forbidden to inoculate from the cow herself?" to which Ring, the Jennerian advocate, replies † that no such prohibition existed. There is no doubt, however, that any reversion to cow-pox, as it occurred in the cow-houses and among the dairy folks, was tacitly discouraged. Jenner himself thus naïvely expressed, some years after, his contentment with the old stock : "If there were a real necessity for renovation [of the stock of lymph] I should not know what to do, for the precautions of the farmers with respect to their horses have driven the cow-pox from their herds." ‡

The Würtemberg collective investigation on cow-pox.— Besides Ceely's evidence on the effects of primary lymph, we have not less candid and accurate observations of somewhat earlier date, by Estlin, of Bristol, and still earlier by Bousquet, of Paris. There are also the results of a system of "collective investigation" in Würtemberg, under the

---

* *An Appeal to the Public; or, the Hazard and Peril of Vaccination, etc.* By John Birch, Surgeon to St. Thomas's Hospital, 3rd Edition. London, 1817.

† *Answer to Mr. Birch, in defence of Vaccination.* London, 1806.

‡ Baron, *loc. cit.*

general direction of Hering, for several years previous to
1839.

The Würtemberg inquiry * differs in its results from all
the rest in the following important conclusion of Hering:
" We find an essential difference between our observations
and the data of Jenner, in respect that the latter describes
the pocks on the udder as passing into phagedenic ulcers,
and regards that character as distinguishing the true cow-
pox from the spurious. . . It is easy to understand that
when the pocks (especially if they be on the teats, or at the
bases of them) are twice or thrice a day lacerated in the
milking, they will take longer to heal than when they are
undisturbed. However, in our observations, there is not a
word said of corroding ulcers, whether in the true cow-pox
or in the anomalous forms." He then quotes the opinion of
Sacco to the effect that " in the cows of Lombardy the cow-
pox is a much milder disease than in those of England, which
have often slow-healing ulcers as a sequel."

As the Würtemberg inquiry has every appearance of
system and comprehensiveness, I have taken the trouble to
go through it carefully, and I shall state briefly the criticism
or appreciation of it that the perusal suggests.

The questions about spontaneous cow-pox in the cow,
about accidental infection of the milkers, and about the
experimental effects of primary lymph on the human subject, were addressed to all and sundry throughout Würtemberg, and the answers were stimulated by an offer of
premiums for approved cases. It would appear also that
there were hints from head-quarters inserted in a popular
almanack upon " what to observe ; " and if these were at all

---

* *Ueber Kuhpocken an Kühen*, Von E. Hering, Stuttgart, 1839, p. 125.

F

the same as in Hering's later circular (18th June, 1838),*
they must be pronounced to be most decidedly of the
"motivirt" or biassed order.

When we come to examine the large number of reports
sent in over a series of years, we find them almost without
exception to be of the most cursory or superficial kind ; compared with such observations as those of Ceely, at the dairy
farms around Aylesbury, they are unauthentic and value-
less. More especially, the observations relate to the cow
or the milker as seen only on one particular day ; there is no
history or sequel of events. The country people of Würtemberg were told to look for vesicles or pustules on the
cow's teats ; and they would appear to have looked for
nothing else, or, at least, to have reported nothing else.
Also, in the group of cases of accidental milkers' infection,
there is not the smallest attempt to give the complete natural history of the disease. It is idle for Hering to suggest
(as Seaton also would have us believe †) that the phagedenic
or indurative ulcerous process had been, in England, *super-
induced* upon or added to the original characters of cow-pox
by the remorseless traction upon the teats in milking the
animal three times a day. Cow-pox had become what we
know it to be by reason of all such circumstances ; had it
not been for the circumstances, it would never have been
known as cow-pox, but would have passed with little or no
notice as an occasional and unimportant eruption of pimples
or vesicles on the teats of heifers, during certain states of
the mammary function, and probably as common, of that
type, in all countries as it was found to be in Würtemberg.
The best proof that the ulcerating and indurating part of

* *Loc. cit.*, p. 168.   † See chapter iv. p. 65.

cow-pox is no mere appendage that may be lopped off, is the fact that on the milkers' hands, and even on their faces, the vesicles pass into the phase of slow-healing ulcers, with a uniformity that is practically decisive for making ulceration the full and unmodified type of cow-pox as a communicable infection.

The Würtemberg inquiry is far too superficial on that, as well as on other points, to have the slightest weight against the observations of Jenner, Ceely, Estlin, Bousquet, and others; and it has certainly no relevancy for the English vaccine of practice. Again, as regards the effects of experimental inoculation with primary lymph, no one case, or series of cases, in the Würtemberg returns is given with such fulness of details as would enable us to form a critical opinion; and, from the practical point of view, it does not appear that stocks of fresh lymph were systematically cultivated (or otherwise than in the way of experiment) from any of the two or three hundred cases of spontaneous eruptions, even for the limited service of the kingdom of Würtemberg itself. As an early instance of "collective investigation," the inquiry directed by Hering exemplifies all the defects of that method, and none of its possible merits.*

Bousquet's vaccinations from the Passy cow.—In no other country than Würtemberg did the search for original sources of vaccine lead to discoveries on the large scale. The Passy cow of 1836 was something of the nature of a wonder, Bousquet sarcastically remarking that the disease

* Instances of original "cow-pox" in the cow would appear still to occur in large numbers in Würtemberg. In the *Mittheilungen aus dem Gesundheitsamte*, Berlin, 1887, (p. 92), the number of such cases in 1883 is stated at no less than thirty. Why is not that source of vaccine utilised?

is as rare in France as it is common in Würtemberg. Attention was drawn to the Passy case by the accident of
vesicles occurring on the milker's hand, fingers, and lip ; it
was from the milker's vesicles that Bousquet and others took
their lymph to experiment with. Bousquet's observations *
on the effects of primary lymph are in close agreement with
those made by Estlin and Ceely two or three years after; in
particular he noticed the prolonged cycle (fall of the scab
about the twenty-fifth or thirtieth day), the extent, depth,
and reticulated surface of the scars, and, still more signifi-
cantly, the not unfrequent occurrence of slow-healing ulceration under the crusts : " I have seen pustules excavate the
skin so deep that they have produced veritable holes in it."
His inoculations were made with matter from the milker's
vesicles or pustules, which were large, semi-globular,
yellowish blebs, without central depression (their subsequent course is not stated). The vaccinations were more
successful at the second remove on the arms of infants than
at the first. His coloured plate shows, in a series of parallel
figures, the differences between the old lymph of 1799 and
the new, more especially the great accession of size after the
eighth day in the vesicles made by the latter, and their protracted course. After remarking on the violence of the
local and constitutional symptoms, and recalling more particularly the case of an infant whose ulcers were long in
healing, he says: " C'est de ce moment que j'ai compris pour
la première fois les frayeurs de Jenner."† In 1840, or four
years after, Dr. G. Gregory found, on a visit to Paris, that
the lymph employed was chiefly that obtained from the

---

* *Sur le Cow-pox, découvert à Passy (près Paris)*, 1836. Paris, 1836.
† *Loc. cit*, p. 21.

Passy milker's hand, although "some of the original matter supplied by Dr. Woodville [in 1799] is also in use."*

Estlin's vaccinations with fresh cow-pox matter from Berkeley, 1838.—In order to show that the English type of cow-pox was somewhat uniform, I shall refer briefly to Estlin's observations made in Gloucestershire in August, 1838, or about the same time as Ceely was studying the disease in Bucks.†

Estlin had been on the look-out for original cow-pox for a number of years; but he could never hear of any. At length word was sent to him of an outbreak at a farm near Berkeley, in August, 1838. The disease was clearly strange to the dairy folks themselves, although they were living in Jenner's own parish; for it was some days before they thought of the cows' udders as the cause of the sores upon their hands.

Twenty-five cows were affected when seen by Estlin, most of them having irregular circular crusts on the teats, while in some the surfaces were raw. All the milkers had sores on their hands in various stages; in one or two persons an eschar only remained; in others soreness still existed. In a boy of thirteen there was a large inflamed vesicle of a yellowish colour between the finger and thumb, and occupying all the space from the third joint of the finger to the second of the thumb. All the milkers had been seriously indisposed, with axillary swelling and tenderness, lumbar pains, and the like. Estlin procured matter from a cow and from the boy-milker's thumb; but the inoculations from both of these sources failed. His stock, which was eventually supplied to many large towns in England, to the Colonies, and to some parts of the Continent, was raised from a girl

* *Trans. Prov. Med. and Surg. Association,* "Report of the Vaccination Section," viii. (1840), p. 88.

† *London Med. Gazette,* xxii., Sept., 1838, p. 977; xxiii., Oct., 1838, p. 115; *ib.* p. 709; xxiv., April, 1839, p. 153; *ib.* p. 968

(Jane——), aged five, who had been domestically inoculated from
a milker's vesicle eleven days before he arrived upon the scene.
From Jane ——, on the eleventh day, he vaccinated a number of
children, only two of whom were infected; in both the development was late, and in one a rash in patches came out on the thirteenth day over the whole body and limbs, being attended with
general illness. At the next remove the areola appeared as early as
the ninth day.

Writing again when he had reached the sixth remove,
Estlin has a number of serious after-effects to record. In
some cases there were cavities under the crusts, "which
would have contained the whole of a pea not of the smallest
size," in two cases there was axillary abscess; in one
case, of a lady re-vaccinated, sloughs formed at the two
places on the arm, and the ulcers were still unhealed at the
end of five weeks. In many of the infants there were
rashes on the skin: "Though the parents have occasionally
expressed uneasiness at these unusual cutaneous accompaniments, they have generally been pleased with the severity
of the complaint."

The next letter brings us to the twentieth remove, at
which stage of the humanising process "the vesicles are
less disposed to be broken during the first week than was
the case at an earlier period." At the twenty-ninth remove
(April, 1839), we have a longer statement about the mitigation of virulence:

"Whether it be dependent upon a more cautious mode of vaccinating [introduction of only a very small quantity of lymph into never
more than two points of insertion], or upon any alteration in the
lymph, violent local irritation and cutaneous eruptions less frequently
accompany the progress of the vesicle at present than was the case
six months ago." However, even at his then writing, "in many cases

the crust becomes rather indented towards the fifteenth day, very like an eschar made with caustic potass, and accompanied by a secondary attack of surrounding inflammation of a more diffused character than the original areola ; the crust is then separated, leaving a small but deep ulcer, that heals in a few days. . . . The only objection to it that I hear of here [Bristol] is its being much more active than the old lymph ; and there are practitioners in other places who, from this cause, have thought it prudent to suspend the employment of it."

The directors of the National Vaccine Establishment made a few trials with Estlin's lymph, the result of which they did not disclose in detail ; but they declined to introduce the new stock into the national establishment, and even hinted that it was " spurious," by which, doubtless, they, as usual, meant to designate something awkward or inconvenient. At Glasgow Estlin's lymph was welcomed,* the public vaccinators there having several times remarked the decadence of the Jennerian lymph. The infants brought back after a week for inspection, on the last occasion before Estlin's lymph was used, presented very poor traces of vesicles ; "in fact, it appeared that in these cases the pock had run its entire course in the time usually allotted to the mere development of the vesicle." The results of the new lymph were very much the same as Estlin's at the corresponding removes :

" In some cases which have been closely under our observation, the constitutional symptoms have come on early, been severe, and seemed to have no relation to the state of the local affection; in a few instances, on the fourteenth day, the spot on the arm has become a deep and angry-looking sore, which has alarmed the friends of the

* " Report of the Committee appointed by the Glasgow Faculty of Medicine," etc., *London Med. Gazette*, xxiv., 1839, p. 208.

child very much; but in none of the cases did the ulceration show any
disposition to extend. Under the application of some mild absorbing
powder, the sore has gradually filled up."

Here, then, we have abundant independent testimony
that the experimental engrafting of primary cow-pox matter
caused the same succession of events as its accidental
inoculation on the hands or faces of milkers. We have,
first of all, the experience of Jenner, which was a good deal
veiled from public view, and explained away by that
experimenter himself according to the doctrine of the
"spurious vaccine vesicle," after he saw the results of
Woodville's lymph. Next in order, but long after in time,
we have Bousquet's experience, in 1836, with primary
lymph from the Passy cow, which gave him for the first
time, although he was director of vaccinations in Paris, an
insight into "les frayeurs de Jenner." Two years later we
have the very precise narratives of Estlin, wherein we may
follow the gradual mitigation of vaccinal effects to the
twentieth, thirtieth, and fortieth removes, the abbreviations
of the cycle and the almost complete elimination of its ulcerative phase. Along with Estlin's own experience we have
a report by a committee of the Glasgow faculty of medicine
upon the effects of lymph sent from the new stock of the
former. Lastly, in order of time, and especially valuable
for the systematic analysis of a wide experience, we have
Ceely's experiments with the primary lymph from more
than one dairy, and the complete elucidation of the same by
his thorough study of the natural history of cow-pox in the
cow herself. All the evidence tends to prove a gradual
mitigation of effects by judicious selection through several
generations of vaccinifers. But, even at some distance from

the source, as Ceely says, "every now and then we have all
the characters of the early removes, and all the inconveniences
of primary lymph." From this we may conclude that the
dangerous fires are still smouldering under the ashes, that
the native wildness of cow-pox is tamed but not extin-
guished, that the virulence is scotched but not killed, and
that it needs only circumstances, or a continuance of
favouring conditions, to bring the dormant characters into
activity again. Keeping this result in mind, let us now
proceed to consider the present-day vaccination practice,
and what are called its anomalies or accidents. Earlier
instances of these misadventures in the ordinary routine
of vaccination have been taken in this chapter somewhat
out of their order by way of illustrating Jenner's doctrine
of the spurious or degenerate vaccine vesicle.

CHAPTER VI.

HUMANISED COW-POX AND ITS ANOMALIES.

It is almost certain that the vaccine matter now in use in
every part of the world is removed by several hundreds
of generations from its parent source and from the character
of primary lymph. Thus we find a private purveyor of
calf-lymph in London * intimating that his stock came from
Rotterdam, and that the Rotterdam stock had been kept
going on the calf's belly for 592 generations, at his then
writing (April, 1881), having been first established by

* Letter in *British Medical Journal*, 23rd of April, 1881.

means of cow-pox matter direct from the Beaugency cow in 1869.

It is obvious from the nature of the references to original cow-pox in recent writings that cases of it are hard to find. Thus in the *Bulletin* of the French Academy of Medicine, for 1882 (p. 17), there is a communication entitled, " Découverte du cow-pox dans la Gironde," in November, 1881, the discovery having been regarded evidently as an event of unusual importance. In England, the editors of the *Veterinarian* inserted a notice in the number for August, 1879, making a request to their readers for lymph " from vesicles * on the teats of cows in cases of so-called natural cow-pox." One result of this request was the intimation, in June following, of a case of cow-pox at Halstead, in Essex ; it was pronounced by Mr. Ceely, who went to see it, to be an eruption of the nature of eczema, whilst Drs. Buchanan and Sanderson, who were present, " expressed no opinion ; " the experiments to produce vaccine vesicles with the matter failed both on the calf and on the human subject.† The only other answer to the notice down to the present time has been a communication from a veterinary surgeon in the west country, to the effect that he had been called two years before (May, 1878) to see two cows with cow-pox : one of them, when seen, was " in the secondary stage," and was doing well ; the other had the teats covered with confluent sores, from which a man on the farm had been inoculated on the finger,

---

* Neither Ceely, nor Estlin, nor Bousquet got matter from " vesicles on the teats of cows ; " they had to use the crusts on the teats or lymph from the vesicular stage of the inoculated sore on the milker's hand.

† *Veterinarian,* September, 1880, p. 597.

and afterwards suffered with "abscesses" of the hand, attended by serious illness for several months.* There is also a comparatively recent reference to an outbreak of cow-pox near Reykjavik, in Iceland, in the summer of 1876 : it had never been seen there before; all the cows at the farm became successively affected; and several of the milkers got inoculated on the hands, much alarm being caused by the severity of the symptoms.†

It would be more curious than useful to trace to its several sources the vaccine matter now in circulation in the various countries of the world. Woodville's lymph held the stage, with few rivals, for nearly forty years, having been adopted by Jenner, and recognised as the true Jennerian vaccine. Its chief rivals, oddly enough, were stocks of horse-grease, or so called "equine lymph," such as those extensively circulated in Italy by Sacco, and in Vienna by De Carro, under the influence of Jenner's original theory. But in both hemispheres it was English lymph from Woodville's stock that was mostly used for many years. That was the case in Paris, as we are told by Bousquet,‡ down to 1836, when that official established a new stock from the Passy cow, without, however, abandoning altogether the old Woodville strain. As regards Great Britain, the Report of the National Vaccine Establishment for 1838 states that they were then using lymph thirty-eight years old, "obtained from Dr. Jenner." It comes

* G. Lewis, *ib.*, October, 1880, p. 695.

† Quoted in *Lancet*, 1880, i. 247, by Fleming; he does not say what kind of sores the cows and the milkers had, but gives a reference for further particulars to the *Deutsche Zeitschrift für Thiermedicin* (December, 1879), a journal which I have been unable to find in the libraries.

‡ *Loc. cit.*, p. 30.

out, however, in the Report of the year after (1839) that the supply of the establishment had "more than once or twice been recruited with fresh genuine matter from the cow," and one of these fresh stocks had doubtless been that discovered by Leese, an officer of the establishment, in 1836. Again, at the Small-pox and Vaccination Hospital of London, a new stock of lymph was introduced in 1837 by Marson from vesicles on the hands and arms of a dairy-maid, "to the ultimate exclusion of the old lymph, whose declining activity Dr. G. Gregory had long noticed and had clearly pointed out." *

The feeling of dissatisfaction with the old lymph was very general in those years, and it seems to have led others besides Estlin and Ceely to seek after cow-pox in country districts.† When the Epidemiological Society made a systematic inquiry into the subject in 1851, evidence was forthcoming of a number of independent or private stocks having been raised in Norfolk, Suffolk, Leicestershire, and other counties.

Whatever new stocks may have been raised during the last quarter of a century, we know as little of the primary disease in the cows from which the matter came, or of the effects of the lymph in its first removes, as we know of the mode in which Badcock's variolous Brighton lymph was "managed," or how Loy's equine matter was mitigated, or what was the history of the "direct equine," which, in Jenner's practice in 1817, gave vesicles "beautifully correct," and was sent to Edinburgh and other places. Some of

---

* Ceely, *Trans. Prov. Med. and Surg. Assoc.*, viii. 357.

† See the Report of the Vaccination Section of the Prov. Med. and Surg. Assoc. in *Transactions*, viii. 19.

these new stocks have certainly had great currency, such as Badcock's * lymph at Brighton, and a corresponding variolous stock raised at Boston, U.S., in 1852.

Besides the certainty that some of the lymph now in use is the variolous matter cultivated by Badcock,† the doubt also arises whether some of the recent stocks, more particularly those used in the remunerative business of calf-lymph establishments, may not have been derived from one or other of the so-called "spurious" forms of cow-pox. When we bear in mind the very vague and generally erroneous notions prevalent as to the nature of the historical cow-pox, or the almost universal assumption that it is a crop of lymph-yielding vesicles on the cow's teats, that doubt has a *primâ facie* warrant. It is no evidence of "genuineness" that the matter can be successfully inoculated : in proof of which statement I shall not enter into questions touching

---

* Mr. Badcock wrote as follows to the *Pall Mall Gazette*, on 23rd Jan., 1880 : "By careful and repeated experiments I produced, by inoculation of the cow with small-pox, a benign lymph of a non-infectious and highly protective character. My lymph has now been in use at Brighton for forty years, and is, at the present time, the principal stock of lymph employed there, being that exclusively used by the public vaccinators." I can find no detailed account of Mr. Badcock's procedure. The attempt in 1836 of Dr. J. C. Martin, of Attleborough, Mass., to "vaccinate" with variolous lymph cultivated on the cow's udder, caused a serious epidemic of small-pox among the "vaccinated" and others. See *Boston Med. and Surg. Journal*, p. 77. Feb. 23, 1860.

† In the official papers of the New South Wales Government, relating to the outbreak of small-pox on board the s.s. *Preussen*, Bremerhaven to Sydney, issued in February, 1887, it transpires incidentally that the English lymph used for re-vaccination at the Quarantine Station of Sydney, was Badcock's. It "took" in a larger proportion of cases than the lymph current in arm-to-arm vaccinations at Sydney, although in a small proportion absolutely.

eczema, impetigo, pemphigus, and the like, but adduce evidence of unimpeachable authority, and relating to an eruption of the cow's udder that is of tolerably common occurrence.

All the writers on cow-pox in the cow have agreed to set aside the white or blister-pock as "spurious." Jenner's reasons for counting it spurious were that it heals quickly under a scab, never eats into the fleshy parts, and is not nearly so infections.* Ceely also reckons the blister-pock spurious; but he upsets the whole of Jenner's reasons. He describes it as "a highly contagious disease among milch cows, and to the milkers, quickly causing vesications and deep ulcerations; often or almost always confounded by them with the true vaccine, and certainly not readily distinguishable in all its stages by better informed persons than milkers." † He showed to Dr. E. C. Seaton drawings of three cases of it on the hands of milkers, "in one of which the appearance of the vesicles singularly resembles that of true vaccine;" and he also showed to that gentleman a drawing of the white or blister-pock on a man experimentally inoculated with it from a milker: "Complete vesicles were formed, with some areola, by the second day; the vesicles and areola were fully developed by the third day, and had then very considerable resemblance to cow-pox at its full or eighth-day development; by the fifth day desiccation had taken place, and the areola was declining." ‡ If we may generalise from a solitary instance, the cycle of the white-pock is a very short one. Full development of the vesicle and areola in three days is a shorter cycle than has been

---

* *Further Observations, ed. cit.,* p. 77.   † *Loc. cit.,* viii. 297.

‡ Seaton, *Handbook of Vaccination,* London, 1868, p. 11 (note).

observed in even the most abbreviated types of vaccination
on the child's arm. It must remain a question whether the
cycle of the blister-pock would not have been much more
protracted if the matter had been taken from a case specially
characterised by the type of "deep ulcerations," whether on
the cow's teats or on the milker's hand. But, taking the
facts as we find them, maturity in three days is within
measurable distance of what happens in ordinary experience
when vaccine lymph is inoculated at many points on the
shaven belly of the calf. In those circumstances, maturity
in four days is not unusual, and maturity in five days is
quite common. It is true that on the child's arm the cycle
becomes longer, just as the constitutional disturbance
becomes much more severe than it is in the calf. But, so
far as the calf's vesicles are themselves concerned, they serve
to show that vaccine matter, from presumably "genuine"
sources, may be brought within a measurable distance of
lymph from the blister-pock, in respect to early maturity.
Thus, one of the last remaining criteria between spurious
and genuine cow-pox presents itself in the aspect of a vanish-
ing difference; and who knows whether the difference may
not have been once and again overlooked?

Considering, then, the variety of stocks of "vaccine"
matter now in circulation all over the world, we shall have
to regard the standard or type of the vaccine vesicle as a
sort of average, which has been attained to by cultivation or
selection, and kept as steady as possible. It is no small
testimony to the adaptability or manageableness of morbid
processes considered as species, that from sources on the
whole so dissimilar as cow-pox, horse-grease, and human
variola, an almost identical type of vesicle should have been

evolved on the infant's arm.   At the same time, for practical purposes, it is only the infection derived from the cow's teats that need be kept in view in this and the following chapters.

The average effects of vaccine in every-day practice are a more or less remote reproduction of the natural history of cow-pox in the cow, of accidental cow-pox in the milker, and of the infection set up by primary lymph experimentally in the child.   The experiments with primary lymph are, indeed, the key to the vaccinal process, and to its so-called anomalies and complications.   However far the vaccine may travel from its source, it can but "drag a lengthening chain," the intermediate links being very conspicuously seen in the earlier removes from the cow.   The statement is no mere theory, but stands upon the very full and clear narratives of facts by Estlin and Ceely.   Since Jenner's first essay, it has been a favourite mode of speaking to describe cow-pox as the small-pox "passed through the system of the cow." If we adopt that phrase as a model we should say, on the evidence of facts, that the vaccinia of every-day experience is the cow-pox modified by passing through the human system.*   Cultivated for a number of generations on the infant's arm, it has acquired the characters by which we know it in ordinary.   But every one of those characters carries us back to the disease on the cow's teats, or on the milker's hand.   Let us take them in order, beginning with the vesicle and scar, proceeding next to the constitutional symptoms, and reserving the areola to the last.

---

* "This animal poison is too mischievous for use until it has been meliorated by passing through some human body, selected as the victim of the experiment."   Birch, in *Letter to Rogers*, ed. cit., p. 137.

The vesicle.—The ideal vesicle is what Jenner, in his later years, used to call "the pearl upon the rose." Its well-known form is due to the fact that the eating away of the tissues proceeds round its periphery both to the depth and to the breadth, causing the soft skin of the child to be raised round the margin into a vesicular pearly ring by the gathering fluid. The corroding process under the skin stops usually about the eighth day; and here we have the first proof that the original process is abbreviated or mitigated.

It was clearly shown by Bousquet, and graphically illustrated by him in a series of parallel figures, that the infection with primary lymph continued its corroding process under the raised skin for several days beyond the point at which the infection with old or humanised lymph stopped. The same was afterwards systematically proved by Ceely and Estlin. These progressive vesicles, continuing to encroach upon the sound margin of tissue up to the twelfth or even the fourteenth day, break at last and become open sores. It is the narrowing of the cycle, or the abbreviation of the process, that saves the vaccinal vesicle, as we ordinarily know it, from that fate; and therein lies by far the most striking part of the mitigation of cow-pox by passing it through a series of infants—a change so remarkable on the surface that the essential similarity is not easy to trace. The whole process, as we now see it after many removes from the cow, takes place under the skin, or under a scab; the vesicular part is retained, and the ulcerous termination left out. In Jenner's words, "it has not passed the boundary of a vesicle." The shortening or contraction, however, has been distributed uniformly over the whole process, so that repair (under the scab) is completed, and the scab itself fallen,

G

within the time that an infection with primary lymph
(accidental in the milker or experimental in the child) would
merely have reached the full limits of its vesiculation.

The scar.—This narrowing of the cycle, and the limitation of it to a subcrustaceous process, has also an effect
upon the character of the scar. The punctated or pitted
scar of what is considered good vaccination is the characteristic scar formed under a crust, where the corium has been
destroyed to some depth. Ceely gives, in his second
memoir,* an account of the appearances of the scars on the
cow's teats: the sores had mostly, if not in every case,
granulated with or without a covering of crusts; the rounded
induration of the margins was still obvious, as well as the
infiltration of the base; and the scars were sometimes
puckered and uneven, but more often regular and smooth.
In his account of the scars after inoculation of the human
arm with primary lymph, he remarks a difference between
the purely subcrustaceous cicatrix where the vesicle had
never burst (the rarer event), and the scar after the healing
of an open yellow foul excavation: the former was a deep
foveolated red cicatrix; the latter a pink, shining, puckered
scar.

The deep foveolated scar is the prototype of a good
vaccination mark, being characteristic of a subcrustaceous
loss of substance; the pitting, like the head of a thimble,
simply means new corium formed in close apposition to the
old scab. However, even after unbroken vesicles on the
child's arm, the marks are not always alike: " On a thick,
sanguine skin," says Ceely, " the cicatrices were deep; but
on a thin skin, shallow; they were not always proportioned

---

* *Loc. cit.,* x. (1842), p. 238.

in width to that of the vesicle, the smallest cicatrix often
succeeding the largest vesicle." Moreover, and this is
important, "after a few months, the state of the arms in
many subjects with thin skins may reveal little of the degree
to which the vaccine influence has been exerted upon them."
The later the crust fell off, of course the deeper the
cicatrix, which on these occasions was often beautifully
striated ; a plump, smooth, and clear skin, associated with
a dark and not too florid complexion, is the most favourable soil for the vaccine matter, and will yield a "magnifi-
cent, well-excavated scar." *

Constitutional symptoms.—We come next to the constitutional disturbance. Its varying severity in the vaccinated infant is a commonplace of practice. It will
hardly ever assume the full force of the disease as it has
been seen in the milkers after accidental infection. It is in
the latter, however, that we have the prototype, the rise of
temperature and the aching of the body at a time when
the seat of inoculation is still a small papule, the axillary
tenderness and swelling (also premonitory), the disordered
stomach and bowels, with delirium now and then.

The exanthem.—Not the least remarkable part of cow-
pox infection is the eruption on the skin during the latter
part of the shortened cycle in the child. In the cow,
according to Ceely,† an eruption consequent upon "vaccine
fever" appears about the ninth or tenth day, in the form
of erythemato-papular elevations of different sizes, from

---

* It is beyond my purpose to show how these anatomical facts in the
natural history of humanised cow-pox bear upon the modern doctrine of
" good " or " bad " vaccination marks.

† *Loc. cit.*, viii. 328.

a mere point up to a vetch, solitary or in groups. In the course of a day or two the papules produce fluid, and at the end of five days will have collapsed, burst, dried, and scabbed; they are mostly confined to the hairless parts; sometimes they come out later than the tenth day, and not unfrequently they continue to form and dry up, and form again and again for three or four weeks. In the accidental infection of the milker, there is in like manner a general eruption at the height or after the decline of the disease,* especially in young persons of florid complexion and sanguine temperament. It may be papular, vesicular, or even pemphigoid. There is very little said, however, of this late eruption in the clinical histories of accidental milkers' cow-pox, these cases having been recorded chiefly with a view to the local sores and the matter got from them. In Woodville's record (39th case) we read that the experiment to retro-vaccinate the cow with matter after two human removes not only succeeded, but that "a man-servant, by milking this cow, was also affected with an extensive tumour upon his thumb: this soon acquired a livid blue colour, and was attended with a considerable degree of fever for several days, and with a rash upon his ankles and feet."† Perhaps the most remarkable case of the kind is one of Ceely's: "Not long since I saw a wife and five children labouring under a *pustular* disease of six weeks' standing, and infected by the father, who had caught the disease [cow-pox] from the cow, which was in a terrible condition. It was of the character of ecthyma, but communicable, affecting the

---

* *Ib. id.*, 337.

† *Reports of a Series of Inoculations, etc.* London, 1799.

face, trunk, and limbs, and could be propagated by inoculation." *

In the systematic vaccinations with lymph of the early removes, Woodville's experience of general eruptions became notorious. Unfortunately it is now impossible to disentangle the cases of true vaccinal exanthem from the preponderating cases of concurrent variolation (or of small-pox accidentally caught at the Inoculation Hospital), Jenner himself having done his best to increase the confusion so as to discredit Woodville's practice as a whole. Ceely is less precise than usual on this head, remarking merely in his section on ",Vaccination of Man with Primary Lymph," that " roseola, lichen, etc., with vomiting, diarrhœa, delirium, etc., arise in some, while in others mere acceleration of pulse is observed, without complaint." In Estlin's first experiment (second remove from the cow) the areola was delayed until the thirteenth day, and at the same time there came out all over the child's body and extremities a rash in patches (maculæ), accompanied by much constitutional disturbance.† At the sixth remove Estlin thus refers to the frequency of rashes : " Though the parents have occasionally expressed uneasiness at these unusual cutaneous accompaniments, they have generally been pleased with the severity of the complaint." ‡ At the twenty-ninth remove he writes that " cutaneous eruptions less frequently accompany the progress of the vesicle at present than was the case six months ago." § The coming

---

* *Loc. cit.*, x. 235 (note).

† *Lond. Med. Gazette*, xxii. (1838) p. 977.

‡ *Ib.*, xxiii. 115.

§ *Ib.*, xxiv. (1839) p. 153.

and going of the exanthem (roseolar, or lichenous, or vesicular) for several weeks is mentioned by several authorities,
including Willan.*

Vaccine roseola, or even pemphigus, has come to be
regarded in a somewhat conventional way, and as if it had
no significance for the true nature of the inoculated
infection, of which it is really the secondary exanthematic
effect.† I shall return to this point in the concluding
chapters, and will be content for the present to quote
a suggestive case of Ceely's, which will at the same time
serve to show the extraordinary "sports" that the systematic
practice of vaccination sometimes brings out : he vaccinated
" a remarkably fine, florid, plump, vivacious infant, aged
eight months, with an active lymph, about eighty removes
from the cow. At the acme of the areolæ of the two
vesicles, nearly the whole surface of the skin of the face,
trunk, and limbs was suddenly covered with large and
elevated erythematous patches and spots, which speedily
became surmounted with vesicles and *pemphigoid* bullæ
of various forms and sizes, exciting considerable and in-
tolerable irritation. But this was not all, for nearly the
whole of the mucous membrane of the lips, cheeks, mouth,
and fauces, as far as the eye could reach, was affected in

---

* *On Vaccine Inoculation* (London, 1806), Appendix vii., p. 41 : a
remarkable case, in the practice of Mr. Farish of Cambridge, of vaccinal
pemphigus which came out time after time ; when the blebs broke, "the
discharge from them inflamed the skin over which it ran."

† Parrot, however (*La Syphilis héréditaire etc.*, 1886, p. 33), remarks
on the resemblance in characters, if not in circumstances, between the
roseola of cow-pox infection and that of syphilitic infection : "Une
éruption qui mérite de nous arreter au point de vue de sa resemblance
avec la syphilide maculeuse, est la roséole vaccinale."

like manner; the whole exhibited a most deplorable sight, and certainly not without danger. Five or six weeks elapsed before the vesicles and bullæ ceased to appear, and the child was restored to comparative health and comfort." *

The areola, and vaccinal erysipelas.—Lastly, we come to speak of the areola, a zone of surface redness round the vesicle, accompanied by infiltration of the deeper tissues, which usually appears about the eighth or ninth day. I have reserved it to the last, because upon it hangs the important question of the relation of erysipelas to the natural history of vaccinal infection. We need have no hesitation in dismissing the theory, which can always be plausibly urged for apologetic purposes, that the erysipelas of vaccination is owing to foul lancets, or extraneous infection introduced. A certain degree of erysipelas was spoken of by Jenner as part of the natural history of inoculated cow-pox; and, indeed, he was at one time not quite sure about its protective power against small-pox unless that, as well as other rather severe symptoms which would be obnoxious to our latter-day sense of what an infant should suffer, was tolerably manifest. Thus in *Further Observations* (p. 135) he says: "In calling the inflammation that is excited by the cow-pox virus erysipelatous, perhaps I may not be critically exact, but it certainly approaches near to it. Now, as the diseased action going forward in the part infected with the virus may undergo different modifications, according to the peculiarities of the constitution on which it is to produce

---

* *Trans. Prov. Med. and Surg. Assocn.*, x. (1842), p. 231 (note); Compare with this Mr. Hutchinson's case of gangrenous eruption after vaccination, *Med. Chir. Trans.*, lxv. (1882), p. 1 ; and Dr. Crocker's cases, *ib. ;* 1887.

its effect, may it not account for the variation which has
been observed ?" (*i.e.* the variation of Woodville's experience in London from that of Jenner in the country.)
Ceely's observations, forty years after, made it probable that
part of the difference, at least, was between the first removes
from the cow and the later.   With regard to the effects
of primary or direct cow-lymph, Ceely emphasises, firstly,
the remarkable redness round the puncture during the first
two or three days (corresponding to the "early erysipelas"
of the Germans), and, secondly, the increased area of the
ordinary areola or rose-red blush round the vesicle about
the eighth day.   It is the latter that constitutes the prototype for the commonest vaccinal erysipelas of ordinary
practice   (the   "late   erysipelas"   of   German   writers).
Ceely's remarks on the effects of primary lymph are : "The
colour and extent of the areola vary, of course, in different
subjects, being very florid and extensive in the sanguine
and irritable, pale and limited in the leuco-phlegmatic and
apathetic; but at its height, and about the decline, there is
considerable induration of the surrounding integuments in
all, influenced by the same circumstances certainly, but
manifestly existing to a greater degree than is observed in
corresponding temperaments from ordinary lymph.   The
areola, under these circumstances, *declines and revives,*
continuing to exhibit a brick-red or purplish hue while the
hardness remains, indicative of deep-seated inflammation
in the corium and subjacent cellular tissue " (*loc. cit.,* viii,
346).

Like all the other so-called anomalies or accidents of every-
day vaccination practice, erysipelas requires a certain rather
unusual concurrence of circumstances to call it forth.   But it

is none the less a latent potency of inoculation with cow-pox matter; and when it does occur, it is to be regarded as a throwing back to one of the original characters of that communicable infection. As Ceely says, the areola, after primary lymph, was manifestly more extensive, deeper in colour (brick-red or purplish), and with more extensive infiltration underneath, than after humanised lymph in children of the same temperament. Lest the evidence of Jenner and Ceely should not seem sufficient for the present day, I shall quote the testimony of Bohn, one of the chief German authorities on vaccination.* After a full discussion of the erysipelatous nature of the normal areola, he concludes : " The lymph of a true Jennerian vesicle, pure and clear, is therefore endowed with a power of engendering erysipelas."

A table of the admitted mortality from "erysipelas after vaccination" is given in chapter ix. The erysipelas engendered in the process of vaccinal infection, or, in other words, by exaggeration of the normal areola and infiltration, may, of course, become the source of erysipelatous contagion to others, just as erysipelas of other origins may so become. In foundling hospitals, particularly that of St. Petersburg, the erysipelas of vaccination has been common, and has spread to the inmates generally.†

An able lay critic of the history and practice of vaccination, who has studied the documents with a closeness of attention hardly to be matched in the profession itself, has recognised the scientific place of erysipelas in the natural history of inoculated cow-pox. I cannot agree with him,

---

* *Handbuch der Vaccination*, Leipzig, 1875, p. 174.

† See the paper " Remarks on Certain Diseases of Infants," by Doepp, of St. Petersburg, translated in the *Lancet*, 1837, vol. i., p. 851.

however, that "the prime note of vaccination is erysipelas." *
The dominant fact of vaccination, in my opinion, is that the
vaccinal process, as we know it, is the contracted cycle of
an infection whose real nature has been almost hidden from
us in the long succession of removes from the cow. But
every now and then, as Ceely says, we have all the characters of the earlier removes, and all the inconveniences of
primary lymph. Erysipelas is only one of these reminders
of the past; and I do not think it is the chief one. Much
nearer to the heart and core of the cow-pox infection lies
the risk that I shall discuss in the next chapter.

CHAPTER VII.

VACCINAL SORES AND VACCINAL SYPHILIS.

"THE origin of the syphilis that occurs as a sequel of
vaccination is shrouded in mystery, and all attempts hitherto
made to penetrate the mystery have failed." These are the
words of Bohn, in his *Handbuch der Vaccination* (1875), at
the end of a recital of recorded epidemics, and a discussion
of their respective circumstances. There are many to whom
such a declaration will come as a surprise, both among the
friends of vaccination and among its opponents. Notwithstanding an overwhelming body of evidence to the contrary,
it is still believed that the virus of the venereal pox may be
conveyed in that of the cow-pox. One of the chief arguments in favour of cultivating vaccine lymph on the calf's

* William White, *Story of a Great Delusion*, London, 1885, p. xxxix.

belly, instead of using the lymph on the child's arm for the inoculation of others, is that the former method avoids the risk of inadvertently transmitting a syphilitic constitutional taint. I shall give the evidence concerning the transmission of syphilis from a syphilitic vaccinifer at the beginning of the next chapter. It has to be said here that it is the entirely negative character of that evidence which explains and justifies the remark of Bohn, that the origin of vaccinal syphilis is shrouded in mystery. But, whether its origin be a mystery or not, the thing itself is no doubt there, a menace and a frequent source of dread to families, and an accident, when it does occur, well calculated to bring the law into discredit with the people. It is with a full sense of responsibility for what I shall have to lead up to in the present and the following chapters, that I approach this concluding part of my inquiry.

"Persons talk very glibly," says Seaton, "about sores being syphilitic, and eruptions being syphilitic, as though the characters of syphilitic sores and syphilitic eruptions were so made out that there could never be any mistake about them. Yet such mistakes are daily being made by practitioners in general, and are occasionally made by the very highest authorities. About four years ago one of those amongst us most conversant with syphilis, Mr. Henry Lee, announced to the Medico-Chirurgical Society that he had a case under his care in which a syphilitic chancre had been produced on the arm of a child by vaccination. The case was seen by many members of the profession, some of whom agreed with Mr. Lee, while others saw nothing but a sore arm, the result of a degenerated vaccine vesicle," an opinion which Mr. Lee himself afterwards came round to. The dispute here was evidently about the name; the morbid condition itself was an anomalous sore, occupying the site of the vaccine vesicle, and it was so like a chancre that a highly competent authority pronounced it to be such. It was a chancre without venereal associations, either direct or indirect; it was such a sore as Jenner speaks of in the "poor girl who produced an ulceration on her lip by frequently holding her finger to her mouth to cool the raging of a cow-pox sore by blowing upon it." Let us, by all means, adopt Seaton's attitude towards "persons who talk very glibly about sores being syphilitic and eruptions being syphilitic;" but let no one be easily satisfied with the formula of "degenerated vaccine vesicle." The term "degenerate" may be used as glibly as the term "syphilitic." To what type does the vaccine vesicle revert when it degenerates; along what road does it travel backwards; is there anything specific in the ulcerous process, or is it merely a common sore arm?

Whoever has read attentively my earlier chapters will admit that a sore with specific characters, whether phagedena or induration or both, is part of the natural history of accidental cow-pox in the milker's hands or face, and in experimental cow-pox set up by primary lymph or by lymph in the proximate removes from the cow. The troublesome nature of these sores, their corroding and indurating properties, their painfulness, their slowness to heal, and the sympathetic enlargement of the nearer lymphatic glands, have been freely admitted by the best authorities. We shall find that these are precisely the characters of the degenerate vaccinal vesicles, which have sometimes been called vaccinal chancres, but more often not called by that name. As "a

small disposition to heal" is John Hunter's broad charac-
terisation of syphilitic sores, it is not surprising that the
vaccinal sores on the arm should have been called syphilitic.
It is still easier to understand why they should have been
called syphilitic when we bear in mind the eruptions and
other constitutional effects that sometimes followed them. We
have seen that roseola, lichen, and even pemphigus, are part
of the natural history of cow-pox infection, just as they are
of infection by venereal sores ; and we have noted also that
remarkable case of Ceely's, at the eightieth remove from the
cow, in which the mucous membrane of the throat was
acutely involved at the same time as the skin. The anomalies of vaccinal syphilis are all explained by the fact, positively ascertained but hitherto disregarded or ignored, that
cow-pox is also a disease with those characters, of inveterate
ulceration and communicability, in which the specificity [of
venereal pox itself consists. There is, of course, nothing
venereal in the cow-pox ; but it has neglect of healing, or
inveteracy with its attendant infectiveness, introduced in a
way that needs only common sense to understand.

It has been remarked by James Moore,* assistant-director
of the National Vaccine Establishment, that "if the cow
could plead her own cause, she might assert that what we
call the vaccine did not originate with her. She might
retort upon us that it was the contact of man which polluted
her pure teats ; for no cow that is allowed to suckle her own
calf untouched by a milker ever has this complaint." Contact of man, no doubt ; but not in the sense that Moore
meant, nor in the sense of Moseley and Birch, or of Jenner

---

* *Reply to the Anti-Vaccinists*, London, 1806, p. 9. Moore is favour-
ably known for his historical writings on small-pox.

himself. The early opponents of vaccination sometimes hinted, although it was not their steadfast opinion, that cow-pox might be the venereal disease conveyed by the hands of men-milkers and dairymaids ; to which there was the very obvious and just reply that the imputation was a slander and an error in fact. It needs the rational analysis of the venereal disease itself to show the way for the rational interpretation of cow-pox.

In a work which I published two years ago, I attempted in one of the chapters to illustrate the antecedents of syphilis as a specific infection, and the acquisition of its specific or autonomous characters, on a basis of neglected or retarded healing of casual sores.* When I was writing that chapter, I was in an average state of ignorance as to the real nature of cow-pox in the cow ; and I had certainly no suspicion that a very occasional and always sporadic disease, arising now and then *de novo* under the circumstances of cow-pox, would illustrate in the closest particulars the same far-reaching effects of retarded healing in common sores which I sought to make good as the rational analysis of the venereal infection.

I now find in the parallelism of the cow-pox with the

* *Illustrations of Unconscious Memory in Disease, including a Theory of Alteratives.* London, 1886. Chapter ix., "The Alterative Cure of Syphilis." The chapter, except in its therapeutic part, is a further application of a pathological principle which I stated originally, and illustrated by some examples, in an address given at the meeting of the British Medical Association in 1883, "On the Autonomous Life of the Specific Infectious," *Brit. Med. Journal*, Aug. 4th, 1883. A highly interesting application of evolutional principles to the pathogenesis of syphilis as a specific disease will be found in Mr. Le Gros Clark's paper on "The Venereal Disease, chiefly in reference to its evolution," *Brit. Med. Journal*, 24th April, 1886.

venereal pox that degree of conviction which arises from the "concurrence of several views in one particular event."*

The earliest instances of sores following vaccination were dealt with in one or other of two ways. In the writings of Jenner and his friends, they were "spurious" vesicles, the spuriousness of the vesicle being merely another way of expressing the inconvenience of the fact. In the writings of the early anti-vaccinists the sores were regarded as an occasional or incidental effect of vaccine according to its proper nature, the child's constitution having a good deal to do with bringing out the effect. Most of the anti-vaccinists within the profession were of opinion that the cow-pox infection was *sui generis*. Thus Moseley † says : "The small-pox is not only destitute of affinity to the cow-pox, but it has no affinity to any other disease whatever. And the small-pox can only be mentioned with the cow-pox to illustrate their mutual dissimilarity. The introducing a bestial humour into the human frame, besides, was not, in my mind, in the most favourable constitutions, a matter of indifference in respect to future health ; and from analogous circumstances I was not without apprehension that, in some habits, the most dreadful consequences might ensue." Again : "The symptoms and demonstrations, whether

---

* The phrase is Hume's, in his profound remarks on the nature of belief : "The concurrence of these several views or glimpses imprints the idea more strongly on the imagination, gives it superior force and vigour, renders its influence on the passions and affections more sensible, and, in a word, begets that reliance or security which constitutes the nature of belief and opinion."—Hume's *Essays*, "Human Understanding," § vi. (Popular Ed., p. 342.)

† *A Treatise on the Lues Bovilla, or Cow-pox.* Preface to 2nd ed., London, 1805, p. vi.

internal or external, of diseases consequential of cow-pox, are totally new, and differ in every particular from established nosological definition " (p. 94).

Birch * represents parents as saying to him that " they are in the most fearful state of suspense, dreading lest what they were persuaded to do in the hopes of saving their children from one disease may not prove the means of plunging them into another, at once novel and malignant." The evil results of vaccination he classified as follows : (1), itchy eruptions; (2) singular ulcerations; and (3) glandular swellings of a nature wholly distinct from scrofula, or any other known glandular disease. Summing up the effects following vaccination in the rather familiar case of " Latchford's child," he says (loc. cit., p. 49), that they "all marked a new and undescribed disease." Once more, he " saw new anomalous eruptions following this disease, eruptions which, in the whole course of his former practice, he had never met with, and must conscientiously refer to this novel practice, and to this alone."

Rogers,† who sometimes represented Birch in the controversy, mentions that the first fatal case which was made public was a patient at Islington, who was seen by Sir W. Blizard and Mr. Cline ; the arm ulcerated, and the patient died. He describes the case of " Latchford's child " as having some resemblance to a case of common boils, only that they returned from time to time. He thus defines the peculiar glandular affection observed in some of the early

---

* An Appeal to the Public on the Hazard and Peril of Vaccination, otherwise Cow-pox, (together with his Serious Reasons.) 3rd ed., London, 1817.

† In Birch, ed. cit., p. 120.

cases: the enlarged glands are "at first the size of a pea, then growing knotty and hard, and at length suppurating." It was "a new disease of the skin, not at all similar to scrofula or any other disease I am acquainted with."

Squirrell was the only one of the more formidable anti-vaccinists who took Jenner's horse-grease hypothesis quite seriously ; and on that theoretical basis he found the evil effects of vaccination to be of the nature of scrofula. One of the most remarkable cases published by him was that of the infant of Smyth Stuart, an eccentric member of the medical profession, who left it for the military service. The infant was vaccinated from a perfectly healthy subject, when it was twenty-two days old, by a very respectable and experienced surgeon in Walworth ; the vesicles did well at first, but on the fourteenth day "inflammation of the arm returned, and extended to a very alarming degree, accompanied with hard painful tumours and blotches, which terminated in obstinate phagedenic sores and ulcers." The child lingered for several months, and the father protested that the only relief it ever got was from mercurial or anti-syphilitic treatment. In his letter detailing the facts to Squirrell, Smyth Stuart wrote : " I was led to consider the cow-pox virus as possessing a suspected venereal taint, or as an infection of the same deleterious quality," the blame being laid on the dairy-folks ; but in the letter as printed by Squirrell, the word " venereal" is suppressed, and " scrofulous " introduced in its place. The wretched state in which this infant died recalls the descriptions of syphilis as it broke upon Europe in the great epidemic at the end of the fifteenth century, rather than the ordinary venereal pox of later times.

II

The first generation of anti-vaccinists died out and left no successors, Birch's tombstone in Rood Lane mutely appealing to posterity to justify his motives, if not his foresight. Sore arms, or "degenerate" vesicles, or "spurious" vesicles, continued as before;[*] and, indeed, whenever vaccination came to be practised on a large scale, these accidents became commonplace. Jenner himself, in his third pamphlet, speaks of the detachment of the scab as "a circumstance not unfrequent among children and working people," and recommends lead lotion to be applied to the part, so as to coagulate the broken surface and prevent a sore. In our own time the frequent use of the "vaccination shield" is an evidence of the risk of some other termination than healing under a scab. The vaccinal ulcer has come, indeed, to be thought lightly of. Thus, Seaton mentions that he once saw a druggist, who had in his time done a good deal of vaccination, about to vaccinate some children with matter taken from an *open sore* on the arm of a child that had been vaccinated on that spot a week before.[†]

Vaccinal ulcers in the early American practice.—The most remarkable instances of ulceration and severe constitutional symptoms on a large scale after vaccination or revaccination come from the United States, first in the period of the early practice in 1800-1802, and again during the Civil War between the Federal and Confederate States. I shall take these incidents in their order.

The record of the earlier disasters will be found in the writings of Dr. Waterhouse, professor of medicine at

---

[*] For example, thirty-five cases of "ruptured vesicles" at the Nottingham Vaccine Institution, *Medical and Physical Journal*, xvi. (1806) p. 137.

[†] *Handbook of Vaccination*, London, 1868, p. 316 (note).

Harvard, who introduced the practice into the New World.*
The first vaccination done in America, with lymph from
Woodville, was upon Dr. Waterhouse's own child, who
suffered from axillary swelling, an efflorescence from the
shoulder to the elbow, and what would seem to have been
an ulcer ; "a piece of true skin was fairly taken out of the
arm by the virus, the part appearing as if eaten out by a
caustic" (*Op. cit.*, i. p. 19). His own subsequent cases were
milder, and in fact regular ; but in the autumn of that
year (1800) a great many misadventures occurred through
the incautious use of vaccine matter from open sores or from
vesicles late in their development. " I have known," says
Waterhouse (ii. p. 8), "the shirt sleeve of a patient, stiff
with the purulent discharge from a foul ulcer, made so by
unskilful management, and full three weeks after vaccination, and in which there could have been none of the specific
virus—I have known this cut up into small strips, and sold
about the country as genuine kine-pock matter coming directly from me.† Several hundred people were inoculated
with this caustic morbid poison." At a later part of his second
essay we come upon the more precise details of these vaccinations with caustic virus : "All those cases where there were
violent inflammations, deep-seated ulceration, eruptions, and

---

* *History of the Variolæ Vaccinæ or Cow-pox.* Part I., Boston, U.S.,
1800 ; Part II., Cambridge, U.S., 1802.

† Precisely the same thing happened to De Carro in Vienna ; and it
was lymph from that source that gave rise, directly or indirectly, to the
fatalities and disasters at Geneva. See Baron's *Life of Jenner*, i. 335.
De Carro, in the first cases, collected the matter "before the pustules
became ulcerated, as it happened by the scratching of my children, which
I never could prevent." Afterwards he used matter which he found "in
great plenty on the sleeves of their shirts."

heavy febrile symptoms were not the true kine-pock, but a
malady generated by a highly acrid, putrid matter ; or, in
one word, poisonous matter taken from under a scab, or
from an open ulcer *long after the specific virus was annihi-
lated.*" The explanation printed in italics is, of course,
sophistical ; the scientific explanation is that the use of
the virus from a late period of the vesicle or ulcer reproduced and gave fixity to that section of the natural history
of cow-pox, which is ordinarily kept latent by careful attention to the period of maturation.* There were two forms
of ulceration clearly distinguished by Waterhouse, the
same two forms that were long after noticed in the
Morbihan epidemic (see p. 139), namely, the indurative,
with rounded sloping edges, and the phagedenic. In the
former case, the vesicle, instead of regularly exsiccating, produced a hard, rough scab after the 11th or 12th day, under
which an ulcer formed, that finally healed by granulations
(ii. p. 97). The phagedenic type, which "may in some be
due to a peculiarity of constitution," is thus described : "At
another time the angry pustule shows no disposition to scab ;
the aperture in the skin increases ; the inflammation blazes
forth afresh, and the illness keeps pace with the progress of
the ulceration ; a transparent glairy fluid fills the cavity,
which granulates very slowly." This transparent fluid had
been used to vaccinate with : " It is the most virulent of all
the discharges of cow-pox. This is the caustic matter which
is apt to produce in patients of certain habits a crop of
eruptions and a heavy weight of constitutional symptoms."

When Jenner heard of the American disasters of the
autumn of 1800 and of the end of 1801, he wrote to

* See chapter vi. p. 97.

Waterhouse that he had been longing for a speaking-trumpet that would carry these words on the rapid wings of the wind across the wide ocean : *Take the virus before the efflorescence appears.* That is, no doubt, the golden rule of safe vaccination. All the same, the disastrous effects of taking late virus, or of allowing vesicles to become ulcers, were neither more nor less than natural and inherent possibilities of all and every inoculation with the products of the disease on the cow's teats. The sophistry of "genuine" and "spurious" vaccine was, and is, quite excusable from the practical point of view of preventing disaster; but, in the natural-history view of the cow-pox as a disease with a definite cycle of potential development, there is neither genuine nor spurious matter, but only early lymph representative of a short and safe cycle, and late lymph representative of a complete and dangerous cycle.

Vaccinal sores, or "spurious vaccinations," among the troops in the American Civil War.—Beside these early American experiences, it will be convenient to place the vaccination disasters of the Civil War sixty years later, by which time the thought of vaccinal syphilis had begun to be entertained. It would take too long to analyse the interesting "Researches upon 'spurious vaccination,' in the Confederate army, 1861-65," by Dr. Joseph Jones, President of the Louisiana State Board of Health.* Many of the bad arms

* See p. 259 of "Circular ii., prepared for the Quarantine Officers and Sanitary Inspectors of the Board of Health of the State of Louisiana." Baton Rouge, 1884. Dr. Jones has collected a great deal of matter, old and new, in his volume or circular; but the book has been mostly left to edit itself. The general look is as nearly that of "shot rubbish" as one may ever expect to see in a book, although the original information is often interesting and valuable.

following vaccination in Confederate soldiers were traced to the "scorbutic condition of the blood." The same evil effects of vaccine lymph among Federal prisoners in Fort Sumpter gave occasion to a formal charge that poisonous vaccine matter had been maliciously used by Confederate medical officers. Phagedenic ulcers, indurations, secondary skin-diseases, and other effects, were produced by vaccination; and these were sometimes referred to "spurious" lymph, acting on an enfeebled constitution, and at other times to lymph taken from a syphilised subject. The evidence as presented to us seems to be a mass of confusion and contradiction, which the most liberal resort to hypothesis and gratuitous assumption can hardly reduce to order. A useful clue to the whole of this gigantic maze will be found to be the natural history of cow-pox, including the latency, but not the extinction, of its ulcerative (phagedenic or indurative) phase. As an example of the perplexity or mystery in which these cases were involved, I quote the following (*loc. cit.*, p. 336) :

"Some of Surgeon Mitchell's cases, which were not fully traced out by him, may have proved syphilitic when fully developed. All those which continued to be under his observation seem not to have been syphilitic. Even this, however, is not perfectly clear. The history of the disease is certainly very suspicious. The patients were previously healthy, and the local results resembled strongly those of syphilitic inoculation. The constitution was evidently involved, but neither secondary nor tertiary symptoms were developed so long as they were under his observation. . . Without his knowledge, Surgeon Brekenridge investigated the disease as it came under his observation, and carefully examined the reports of surgeons transmitted to him in compliance with orders. He concludes : 'that the disease was essentially syphiloid, and in respect as a whole resembled in its incipiency,

progress, or termination, the genuine vaccine disease [paradoxical as it
may appear]. . . There was some tendency to scurvy [among the
troops], but no connection could be traced between the disease and this
condition. . . There was no case in which I had reason to believe
that any antecedent constitutional vice, either inherited or acquired,
exercised the slightest influence in developing or modifying the
disease.' "

The earlier series of disasters in America were fairly
traced to the cow-pox, with the proviso, indeed, that it had
lost its specific properties and acquired acrid or caustic properties (not less " specific," it may be said), but without any
attempt to bring in syphilis as a complicating factor. In
the later series, during the Civil War, syphilis was alleged
by some and denied by others; and, as the quotation will
have shown, the whole class of incidents was honestly felt
to be paradoxical or mysterious.

These events on the large scale have had their counter-
part in every-day practice. On the one hand, there is the
common " vaccinal ulcer," and on the other hand, there are
cases in which it is thought necessary to seek high and low
for a source of syphilitic contamination. The distinction, I
venture to say, is arbitrary; or, at the most, it is a difference
in degree, and not in kind, and a difference between sporadic
cases, taken as a matter of course, and groups of cases apt
to create a stir and to ensure inquiry. The following is
Bohn's description of the " vaccinal ulcer" of ordinary
practice : *

" The destruction of the corium extends both to the breadth and to
the depth, and a crater-like sore mostly results, with a hard base and

* *Handbuch der Vaccination*, Leipzig, 1875, p. 166. Seaton passes over
the subject with a few words.

indurated edges, which, at the first glance, may frighten the practitioner by its likeness to syphilis. The sore is of a sluggish nature,
having little innate disposition to heal. Often there springs from its
floor a growth of spongy tissue, in which case we have weeping ulcera
elevata, with still less of spontaneous disposition to heal. Usually,
only one or two of the vesicles on an arm go wrong, the others scabbing
correctly. The accident is most apt to happen in the warm months of
summer, or when several vesicles are close together, or when the scari-
fications have been made long and deep. White precipitate ointment or
blue-stone lotion will make the sores to close ; the scars are permanent,
and are distinguished by their size and their irregular, lumpy surface."

Bohn adds the perfectly gratuitous statement that these
ulcers owe their existence, *always* and *exclusively*, to some
noxious influence from without ; their disposition towards
phagedena, for example, has nothing to do, he says, with
their vaccinal origin.   This is a good sample of the disincli-
nation of otherwise competent observers to face fairly the
natural-history facts of cow-pox, as they may be read in the
authoritative writings of Ceely, and of Jenner himself.
The vaccinal ulcer is neither more nor less than a reversion
to the original type and full cycle of cow-pox as it occurs,
or used to occur, on the cow's teats, on the milker's hands
or face, and on the child's arm after vaccination with
" primary lymph," or with lymph of the first removes.

The vaccinal sores which did at length raise the question
of vaccinal syphilis were in no wise different from the
spreading ulcers with hard base and indurated edges as above
described.   They raised the question of syphilis communicated along with the vaccine, because they occurred in
groups ; a large number of children vaccinated with the
same lymph on the same or succeeding days, or of adults
re-vaccinated, experienced the same effects ; the community

of effect called attention to the matter in a way that an isolated case would not have done. These disasters were first
noticed in Italy, France, and Germany, about 1830 or
earlier. The first impulse was to accuse the vaccinator of
having used syphilised lymph; and, indeed, two or three
vaccinators were tried in France and Germany on a criminal
charge for that offence, under the codes of those countries.
They were either acquitted, or subjected to a nominal
punishment, for the reason that there was nothing in the
constitution of the vaccinifer to warn them of danger likely
to ensue, the lymph that they used having been taken from
healthy infants, and from vesicles to all appearance correct.
These trials gave rise to much comment and inquiry on the
Continent; and a very important general law was brought
to light, which has, unfortunately, not been kept steadfastly
in view. I shall give it the first place in a new chapter.

CHAPTER VIII.

VACCINAL SYPHILIS.

THE experiments of Bidart in 1831, of Taupin and others
in 1839, and of various members of the Medical Society of
Vienna, subsequent to the celebrated Hübner trial in 1852,
have proved beyond all question that there is no difference,
cæteris paribus, between the vaccinal vesicle of a syphilitic
child and of other children, and that lymph from a syphilitic

vaccinifer, if it be taken at the usual safe stage of maturity, will produce a correct vesicle and not produce syphilis.* The presence or absence of constitutional syphilis in the child was thus shown to be irrelevant for the course of the vaccine infection. This conclusion was at one time welcomed as disposing of the allegations that syphilis had been communicated, as a matter of fact, by vaccination. † Unfortunately something very like syphilis had, as a matter of fact, ensued from vaccination ; and the same unfortunate consequences of cow-pox inoculation have continued to occur from time to time.

As the fact could no longer be ignored, the theory was started that it was not the lymph of the vesicle, but the child's syphilitic blood drawn in the act of taking lymph, that conveyed the constitutional taint to the vaccinated child. This was Viennois' celebrated hypothesis of " vaccine by the lymph, syphilis by the blood," which was debated at inordinate length in the Académie de Médecine ‡ in 1861— 1864. According to Bohn's summary of the evidence, the question of transmission by the blood in vaccinating is now decided in the negative sense. §

One more attempt ‖ was made (in Germany) to uphold

---

* On the other hand, it was proved by Auzias-Turenne that the lymph of a syphilitic child, which was safe at the eighth day, was dangerous at the eleventh. (*Gaz. hebdom.*, 3rd Feb., 1865.)

† See Simon's *Papers relating to the History and Practice of Vaccination*. London, 1857, pp. lxiv.—lxvii.

‡ See the collection entitled *De la syphilis vaccinale*, Paris, 1865, which contains memoirs by Viennois, Pellizzari and others, on particular outbreaks.

§ *Handbuch der Vaccination*, p. 335.

‖ See Bohn, *loc. cit.*, p. 335.

the doctrine of a dual and simultaneous transmission, on the ground that, although the lymph of the vesicle was unable to convey syphilis, yet its base or floor might be so indurated in a syphilitic child that a syphilitic virus or secretion would be produced therein, and might be extracted by a deep puncture. A subtlety of that kind serves merely to show the straits to which scientific medicine was driven for an explanation. Even if it were not a fallacious mode of arguing, it would be of no use for those cases (certainly the majority, if not the whole) where the vaccinifer has not been syphilitic.

No other theory of dual transmission would seem to have been attempted since that time.

It is clear that such a case as the recent rather notorious English one,* full of ambiguities as it is, cannot

---

* I shall refer here briefly to the experiment of Dr. Cory, so that I may not seem to have overlooked a piece of evidence that has lately been made more of by the English profession than the numberless experiments of Taupin, Sigmund, and others, forty or fifty years ago. That gentleman had been three times vaccinated successfully in the ordinary course. In 1877 or 1878 he again vaccinated himself, this time from a syphilitic child's vesicles, and once more with the correct result. On the 5th Nov., 1879, and on 11th May, 1881, he repeated the attempt to raise vesicles on himself from the arms of syphilitic children, but failed; and on 6th July, 1881, he tried for the last time. On this occasion the child was about three months old, syphilitic, and presenting an eruption on its arms; it had five vaccine vesicles, shallow, and difficult to prick without drawing blood. Oddly enough, a most essential fact in the case, the date of the infant's vaccination, or the age or maturity (whether backward, or too early, or average) of the vesicles when lymph was taken from them on 6th July, is not stated in the report upon the case prepared for the medical officer of the Local Government Board (Report for 1882, Appendix No. 7, p. 46). The description suggests that they were backward vesicles, such as have often proved dangerous whether the vaccinifer were syphilitic or not; the

weigh in the scale against the mass of testimony that
syphilis of the child is, *cæteris paribus*, irrelevant for the
course of its own vesicles, or for those of persons vaccinated
from it. The relevant things are the presence of an eruption of any kind (even itch) on the vaccinifer, the retarda-
tion of its vesicles thereby caused, and the use of such
backward and scanty lymph for vaccinating with. It
needed no experiment to prove that anomalous vesicles,
and even vaccinal ulcers, might follow under such circumstances.

The origin of vaccinal syphilis remains, as Bohn says,
"shrouded in mystery." Readers who have followed my
argument hitherto will not be surprised if now I claim the
phenomena of so-called vaccinal " syphilis " as in no respect

eruption on the child's skin could hardly have permitted them to be otherwise, and we are told that there was not enough lymph in the remaining
three vesicles, not used for the experiment, to charge a single tube with.
The first vesicle that was opened yielded mostly blood, and the lancet so
charged was not used ; at the second attempt on a new vesicle, a head of
lymph was obtained without squeezing, but only a small one. With the
lancet so charged, three punctures were made on the skin below the bend
of the elbow. By the 21st day (July 26th) two of these spots had become
red, and had developed small pimples, which grew slowly at first. The
lower one remained papular throughout, until it was excised on 11th Aug. ;
the upper one, on 4th Aug., disengaged a scab from its centre, and appeared to be slightly moist beneath. On the 8th it showed a little yellow
spot in its centre, which was a scab next morning ; on the 11th Aug. it
was still covered by a very small scab, which, when removed, revealed a
little ulcer ; the same day the papules were pronounced by several physi-
cians and surgeons to be syphilitic, and were excised. A slight areola had
appeared intermittently round each of them during their progress. Mercurial treatment was begun after the excision ; on the 31st Aug. a roseolar
rash came out on the forehead, etc., and lasted four days. The history is not
carried farther in the report, nor in any other document known to me ;
but other symptoms are understood to have followed.

of venereal origin, but as due to the inherent, although mostly dormant, *natural-history characters of cow-pox itself.*

With a view to discover the more general circumstances under which so-called vaccinal syphilis has occurred in groups of cases, and to show the small reason in fact, or total want of reason, for assuming the contamination of the lymph by venereal syphilis, it will be necessary to enter somewhat fully into details.

So-called unauthentic cases of vaccinal syphilis.—Of all the cases put on record since the first Italian epidemic in 1814 at Udine (reported by Marcolini), a considerable number have been set aside as unauthentic. The rigorous scepticism with which the allegations of epidemic syphilis due to vaccination have been received, is not surprising when we bear in mind that a conveyance of syphilis by vaccine lymph has been shown by hundreds of experiments to be highly improbable, if not absolutely impossible ; and, secondly, that the inculpated vaccine matter could hardly ever be traced to a syphilitic constitution of the vaccinifer. It is still less surprising that isolated cases of indurated or phagedenic sores at the seat of vaccination (attended by secondary symptoms) should either have been ignored altogether, or summarily dismissed as due to pre-existing but hitherto latent syphilis in the vaccinated infant, although the hypothesis was an imputation on the parents, which, in most cases, they might very justly have resented. So far as isolated cases are concerned, that is the conventional way of disposing of them still. The unwillingness of the profession to accept even the facts of these post-vaccination disasters is well shown by the reception given to a series of cases published in 1859 by Dr. James Whitehead of Manchester.* Whoever takes the trouble to read Dr. Whitehead's observations at first hand will, I think, agree with me that they bear the marks of good sense and reasonableness. The children brought to the hospital for whatever complaint were systematically examined (or their parents questioned) as to vaccination, and 1,435 out of 1,717 were found to have been vaccinated.

"In a considerable number of instances," he says, " the mothers inculpated vaccination as the cause of the diseases under which the children laboured ; but in a certain proportion of these, after patient investigation, no satisfactory grounds could be obtained to substantiate that imputation. In thirty-four of the inculpated cases, however, the evidence appeared sufficiently convincing to warrant the belief that a taint had been communicated ; and in fourteen of these the disease thus implanted was of a true syphilitic character, as the nature of the symptoms and the mode of its derivation convincingly demonstrated. In the remaining twenty cases, whose whole history was less clear, the symptoms in the child were so precisely like those of constitutional syphilis, and so unlike, in several of their features, any other form of disease, that the treatment employed was that commonly used in syphilitic disease, and in most cases was attended with satisfactory results."

From this it will be seen that Dr. Whitehead himself excluded from his diagnosis of syphilis twenty of the cases, apparently for no other reason than that the history was obscure ; it is impossible, however, to go back upon the facts, as these cases are not tabulated with the rest. His table of undoubted syphilitic cases contains sixty-three, of

* *Third Report of the Clinical Hospital, Manchester.* By James Whitehead, M.D., London, 1859, p. 51, and table of syphilitic cases.

which fourteen are put down as due to vaccination. In all the fourteen the parents are acquitted of syphilis : while the health of the vaccinifer was probably not investigated, and is not referred to. In six of them the vaccinal vesicles, or scars, became indurated or angry sores ; in most of those for which primary ulceration is not stated, many months had elapsed before the children were brought to the hospital, and the induration of the scars or ulceration of the vesicles may not have been easy to ascertain by testimony ; in only two out of the fourteen might there be some reason for very rigid scepticism refusing to accept the author's view of the sequence of events. None the less, as Seaton says, " very little weight, I believe, has ever been attached to them." * It is clear that very little weight was attached to these cases, because they fitted in with no one's then views of what was possible or credible.

Not only in isolated cases, but even in groups of cases where the syphilis befel a number of infants vaccinated from a common source, doubts have been thrown upon the authenticity of the facts, just because that common source could not be shown to have been tainted with the virus of syphilis. Thus, Seaton hesitates to receive as authentic the Italian endemic of 1821 near Cremona, because it was not proved, nor even alleged, that the child from whom the inculpated lymph was taken had ever had syphilis. Again, in the second Italian endemic (1841), also reported by Cerioli, wherein sixty-four children were vaccinated, with disastrous consequences, from one child, the latter never had syphilis, and, therefore, the facts, as a whole, came somehow to be set aside. Another suggestive epidemic, published

* *Op. cit.*, p. 326 (note).

under the name of spurious vaccinal syphilis,* in 1870, or long after the possibility of syphilis, due to vaccination, had been admitted, was as follows :

At Argenta, near Ferrara, in September, 1866, vaccine sent from a distance in tubes was used on the 25th to inoculate a healthy infant of seven months ; three regular vesicles formed, there was no unusual disturbance of health, and the child was found to be quite well when examined seven weeks after. On the 30th (sixth day), lymph was taken from its vesicles and inoculated on seven others, who also did well. From the latter, or from one or more of them, vaccine was obtained to inoculate thirty-four children, of whom all but seven developed ulcerated arms. The ulceration is said to have begun from the fourth to the tenth day after the insertion of the matter, indicating prematurity ; elsewhere it is said that in some the vesicle changed into an ulcer, while in others an ulcer developed under the crust ; some of the ulcers became phagedenic, and several were covered with a diphtheritic deposit. In every case they healed without treatment in six or eight weeks. Only six were reported to have had an eruption on the skin ; the condition of the lymphatic glands escaped attention. From one of the infants with ulcerated vesicles, matter was taken to vaccinate seven others, and " iu two of these the vesicles ran a normal course notwithstanding," from which we may infer that in the other five something else happened.

Although this severe epidemic was published in France (some three years after its occurrence), under the title of "Spurious Vaccinal Syphilis," it differs from those that we shall come to in the sequel in no essential respects.

On the same grounds we ought to reject very nearly every epidemic, or group of cases, that has ever been traced to vaccination from a common source ; for, not only have we the experimental improbability, but the common vaccinifer has, as a matter of fact, either been free from syphilis from

* Gamberini, in *Gazette des Hôpitaux* (1870), p. 505.

first to last, or has only been found with some trace of secondary symptoms several months after, which were much more likely to have been the concurrent effects of its own vaccination. Having gone over all the groups of cases or epidemic outbreaks, I see no reason to place any of them in a " spurious " class or to raise the question of authenticity. They must all have been real enough to the poor people themselves; and they are complete in everything except the necessary passport to the sphere of our belief, namely, a consistent theory.

For brevity's sake, however, I shall omit the Udine epidemic of 1814, the two endemics recorded by Cerioli (1821 and 1841), the Lupara endemic of 1856 (in which the search for the syphilitic source was a very late afterthought, and was not in any sense successful), and the endemic of 1862 at Torre de Busi, near Bergamo, which was started by a child with an eruption (and therefore with backward vesicles), but like the other epidemics had no ascertainable origin in syphilis. In all these, the accidents that befel the vaccine vesicles or scars, together with the secondary symptoms and other after-effects, were practically the same as will be described in detail for other outbreaks. The cases among the troops in the American Civil War I have taken in the previous chapter, so as to place them alongside of the very similar American cases in the first years of the century, for which syphilitic contamination was not seriously thought of as the cause. Apart from these omissions or transpositions, the instances given in the sequel do not exhaust the list of epidemic vaccinal disasters (I make no attempt to enumerate the isolated cases). In particular I may refer to the omission of the Hübner case, in Upper

I

Franconia (16th June, 1852), which had ambiguous elements, and was the subject of much discussion for some time after ;* and of the severe village endemic of Dipson, near Pesth, from 1855 to 1857, for which a far-fetched source of syphilitic contamination was discovered in a sore contracted on the forearm by the vaccinifer's grandmother in the course of her duties as a midwife.†

Some epidemics of vaccinal syphilis analysed.—In the Coblenz case‡ (1849) twenty-six persons, mostly adults, were re-vaccinated by a surgeon of the second grade (known in the French discussions as ' le vétérinaire B——'), the lymph having been taken on February 4th direct from the vesicles of a well-grown and apparently healthy child of four months. Seven other children had been vaccinated with the same lymph and at the same time as the inculpated child ; but in none of these other cases did the vesicles run an unusual course, or lead to further consequences. The peculiarity in the child who was the source of the disasters was that its vesicles were not ripe at the usual time (eighth day) ; the intended vaccinations from its arm were therefore put off until the eleventh day, when seven persons were vaccinated, and again until the twelfth day, when nineteen were vaccinated. These were all that the official inquiry recognised ; but the surgeon declared that there were still others. The inoculation held in nearly all the twenty-six, and the vesicles pursued a regular course. However, about three or four weeks from the insertion of the matter, the scars or crusts opened in two of the seven vaccinated on the eleventh day, and in all the nineteen vaccinated on the twelfth day ; specific ulcers

* *Intelligenzblatt der bayr. Aerzte* for 1854.
† *Oester. Zeitschr. für prakt. Heilkunde*, 1862 ; Bohn, *loc. cit.*, p. 322.
‡ The incident is reported by Wegeler in the *Preuss. Vereins-Zeitung* (1850), No. 14. I have not succeeded in finding this periodical in libraries, and have had to depend upon the abstracts of the original paper in Schmidt's *Jahrbücher*, vol. lxvii. (1852), p. 62, and in Bohn's *Handbuch*, p. 313. Seaton has clearly been misled in some particulars through trusting to Depaul's version of the case.

ensued and constitutional symptoms (cutaneous eruptions, etc.), which were treated with mercury. The vaccinated belonged to various ranks of society. The course of the vaccine vesicles on the vaccinifer's arm is not recorded beyond the date of taking matter from them ; but the child was attacked with ‘ water on the brain ’ on the eighteenth day, and in two days was dead, having developed subsequently to the twelfth day an eruption on the inside of the thighs, on the buttocks, and on the face. The surgeon was fined, and imprisoned for two months.

In this case the vaccinifer at the age of four months was vigorous, and seemingly free from disease ; no other aspersion could be cast on the parents except that the child was born out of wedlock. Its vesicles came to late maturity (eleventh or twelfth day), it had a general eruption several days after, and died on the 20th day with brain symptoms. As regards those vaccinated from it, the lateness of the vaccine was very clearly shown, and was admitted by the court, to have been an essential factor in the anomalous course of the infection, only two out of seven vaccinated with eleventh-day lymph having developed the symptoms, while the whole nineteen vaccinated with twelfth - day lymph were affected.

In the Rivalta case * lymph was sent in a capillary tube from Acqui, a town in Piedmont, to the neighbouring large village of Rivalta, where there seems to have been no resident medical practitioner. On the 24th May, 1861, a child, Chiabrera, was vaccinated with the lymph, and developed vesicles in due course. On the 2nd June (tenth day) forty-six children were vaccinated from Chiabrera's vesicles ; and on the 12th June (again the tenth day), one of these served as the vaccinifer for seventeen children more. In thirty-nine of the first series, and in seven of the second, ulcers developed

---

* Reported by Pacchiotti, “ Sifilide trasmessa per mezzo della Vaccinazione in Rivalta presso Acqui.” Torino, 1862.

in the inoculated spots at various intervals from the tenth day up to
the end of the second month, and the specific disease was communicated by contact to the mothers and other persons in the village,
until the total affected reached the number of seventy-eight. The incident passed without notice beyond the village until four months
after, when word of it reached Turin, and a medical commission was
sent to inquire. They found that seven of the children had died, and
in the survivors they found either open sores on the arms, or papules
(? warty excrescences), or scars, sometimes blanched, but more often
copper-coloured. The child Chiabrera, vaccinated with the Acqui
lymph, and the direct vaccinifer of the first forty-six, was found, at
the end of four months, to be ailing, to have lost his hair, and to be
suffering from an excoriated tubercle on the foreskin ; a few months
later his health was excellent. His mother had an ulcer on one nipple,
caught from the child, and a scar on the other ; a few months
later (January, 1862) she had mucous tubercles (*apud vulvam*)
The father was perfectly healthy. The other vaccinifer, who
furnished lymph for the second series of seventeen, was dead one
month before the commission reached Rivalta ; the child was reported
to have developed ulcers at the inoculated spots, a general eruption, and
mucous tubercles near the mouth and *circa genitalia*. The person who
performed the vaccinations was acquitted of all blame ; but some suspicion was thrown on the child Chiabrera, because he had been
suckled, two or three months before his vaccination, by a young woman
who had been syphilitic *for a year and a half*, and had lost her own
child, a theory of the events evidently set up for want of a better, and
entirely unsupported by proof of actual syphilitic infection of the
child by his temporary nurse.

A severe epidemic, not unlike that of Rivalta, occurred
in 1870 in two parishes of the Austrian province of Styria.[*]

The vaccine lymph was sent from Vienna by a practitioner in
private practice, who had no accidents with the stock himself ; the
same lymph was used by the Styrian vaccinator in other parishes of

---

[*] Reported by Kochevar, in the *Allgem. Wiener Med. Zeitung*, 1870
(Nos. 21 and 24) ; abstract in *Archiv für Dermatologie und Syphilis*, 1870.

that province, also without accidents. But in the parish of Schleinitz he vaccinated with it a child, of whose condition nothing special was noted at the time, and who became, on 6th July, 1869, the vaccinifer of thirty-six others in that parish, and of four in the parish of St. Veit. On the 30th November, when an inquiry was held, the vaccinifer (vaccinated with Vienna lymph) was found to be well nourished and strong for her age, of a good colour, but with an ulcerated mucous *plaque* on the right labium, and another in the right groin, as well as several small sores *circa anum ;* the state of her vaccination marks is not mentioned. These effects seem to have followed vaccination at an uncertain interval ; it is positively stated that the infant had no ulceration or condylomata previous to that operation. As regards the thirty-six healthy children vaccinated from this child in the same parish, and the four in the parish of St. Veit, three of the former and two of the latter did not develop vaccine vesicles at the place of insertion of the matter, and therefore had no syphilis, local or constitutional. In all the rest, save one, papules and vesicles developed in due course, which were reported by the mothers, in the retrospect, to have been filled with 'impure' serum. The vesicles broke, and brown crusts formed, covering ulcerations of a dirty whitish colour, which afterwards showed ordinary whitish scars. The glands of the arm-pit and neck swelled, pustular eruptions came out over the children's bodies, and about six, eight, or ten weeks after the vaccination there were condylomata, or sharply-cut ulcers (*apud genitalia necnon circa anum*), with psoriasis, or some other rash, on the skin, and whitish ulcerations about the angles of the mouth. Most of the patients wasted, lost their hair, and suffered more or less from hoarseness and deafness. Several of them died, but it does not appear whether death may not have been due to some intercurrent disease. As in the Rivalta case, several of the mothers and other members of the households were affected through contagion, of whom fourteen were treated at the General Hospital of Graz.

Side by side with the Styrian series of cases we may take Mr. Hutchinson's London cases * (1871 and 1873),

* *Med.-Chir. Trans.*, liv. (1871) and lvi. (1873).

which were the first to rouse general attention to the subject in this country, Whitehead's cases of 1859 having been ignored.

In the first series, twelve persons were successfully vaccinated with lymph taken on the 8th day from a specially healthy-looking child with five good vesicles. When this child was examined two months after, it was found to have correct scars; but it had five small condylomata *circa anum*. There was no imputation on the soundness of its parents. Of the twelve persons successfully vaccinated from the child's arm (most of them in three places, some in four), the first two had no ill effects, but in each of the remaining ten the scars broke out after having come to rest in the correct manner, and in the 8th week presented the appearance of indurated chancres. Under mercurial treatment the induration soon became soft, and the sores healed. Besides headache in some, there was hardly any constitutional disturbance while the sores were present; only two (Nos. 4 and 5) had ulcerated tonsils, and not more than half had a well-marked secondary eruption on the skin.

This series had hardly begun to be talked of, when one of Mr. Hutchinson's hospital colleagues (Mr. Warren Tay) came upon traces of another series; and the following facts were elicited, much to the surprise of the public vaccinator and general practitioner concerned:

Two children of the same family, one aged four years and the other sixteen months, had been vaccinated seven weeks before they came to be treated for skin eruption; and, when the arms were looked at, the vaccination spots in one child were unhealed and indurated at the base, while in the other the scars were unbroken but indurated. By means of the Vaccination Register, twenty-six others vaccinated with the same lymph were traced, and nine of these were found to have chancres on their arms; they had all been vaccinated in more than five places, several of them had merely the local sores, while others had a papular scaly eruption and other secondary effects (ulcers on tonsils, etc.). The vaccinifer, when seen three months after its vaccination, was found to be a stout, well-grown child of seven months; its vaccine vesicles were reported to have done well, and it had correct marks. It had, however, a small condylomatous patch, *circa anum*, in process of healing. The father and mother looked healthy, and the former, when questioned, positively denied having had syphilis.

Mr. Hutchinson's other cases are of two years later date.[*]

The first was a man aged 46, suffering from iritis. Inquiry having been made as to vaccination, he said that he had been vaccinated three months before, at the same time as his three children, who took no harm. The vaccinator, on being appealed to, said that he had inoculated about a dozen more with the same lymph, and that only two or three of them had had a little trouble with their arms; in this one patient, however, the spots ulcerated so much in the manner of phagedena, that he twice applied a strong solution of nitric acid to them, although it did not occur to him to regard them as syphilitic chancres. When Mr. Hutchinson saw the patient for iritis three months after vaccination, the vaccination spots were ulcers as large as shilling-pieces, covered with scabs, and with dusky indurated borders; there was also an indolent swelling of the axillary glands, a papular scaly rash on the skin, and symmetrical ulcers on the tonsils. The ulcerations of the arm dated from the fourth week after vaccination, and were attributed by the patient to the irritant effect of the dust of tobacco, in which he worked; the skin eruption appeared about the sixth week, and the iritis a month later. He was treated with mercury and got well.

The child who furnished the lymph for this man, as well as for three members of his family, and for about a dozen more, was found to be a large fat baby, with no rash, condylomata, or other suspicious features, unless, indeed, a broad and somewhat sunken bridge of the nose have a worse than Shandean import. Its parents were healthy,

---

[*] *Med.-Chir. Trans.*, lvi. (1873).

The last of Mr. Hutchinson's cases was a female private patient aged forty-six.

She had been revaccinated at the same time as her two daughters, neither of whom had any ill effects. In her own case the scars reopened a month after healing, and continued for three months in the form of large ulcers with hard edges; she had also severe and protracted constitutional symptoms. The child from whom the lymph was taken had sores *circa anum*, when seen by Mr. Hutchinson a good many months after its vaccination, and was reported by its ordinary medical attendant to have been treated for syphilitic condylomata; but the sequence of events in the child's case is altogether beyond unravelling.

In three of these series of cases, a diligent search discovered condylomata *circa anum* in the vaccinifer; but that condition was certainly later than the vaccination, and, as in the Styrian outbreak, may not unreasonably be taken as an effect in the vaccinifer concurrent with and equivalent to the post-vaccinal disease in those vaccinated from it. In the other series there was no evidence of any disease in the vaccinifer. In the parents of none of the four vaccinifers was there any admission or even well-grounded imputation of syphilis. In the first series, two out of the twelve vaccinated had no ill-effects; in the second series, only eleven out of twenty-five could be shown to have taken harm from vaccination, and of these several had merely ulcerated arms, which would have passed as commonplace events but for the general inquiry raised; in the third series only one person (an adult) out of twelve or fifteen felt the bad effects of vaccination severely (two or three more having had "a little trouble with their arms"), and in his case the mischief began with a phagedenic tendency in the vaccinated spots;

in the fourth series, the mother only was affected, her two daughters vaccinated at the same time having escaped the risk.

Some ten or twelve years before the date of these London cases, the question of vaccinal syphilis had begun to be debated in France, and more particularly at the Paris Academy of Medicine, whose " Bulletin " for a succession of years, previous to 1869, contains the reports of discussions, as well as the records of several serious outbreaks. Any one who reads these debates cannot fail to be struck by the general sense of perplexity among the members ; at first a majority declared against the existence of vaccinal syphilis, but two years later the Academy yielded to the insistence of the minority, and came to a unanimous opinion in its favour, although the incidents unquestionably remained a riddle to all parties. The following instance will serve to show the nature and circumstances of the disease : *

On 20th May, 1866, an experienced midwife of Granchamp, near Vannes (Brittany), who held two silver medals for vaccination, received vaccine lymph *sur plaque* from the Prefecture. Next day she vaccinated, at two or three places on each arm, two healthy infants named Mahé and Norcy ; and from the latter she took lymph on the eighth day, and vaccinated Marie Rosnaro, aged three months. As this infant was destined to be the vaccinifer of a very large number, she was vaccinated at six places on each arm ; all the twelve vesicles rose and developed correctly. On the 5th of June (being the ninth day) the infant was carried from village to village in attendance on the

* *Bulletin de l'Acad. de Méd.*, xxxii. (1866-67) p. 201 and p. 1,033. The first reference is to a long and somewhat pointless official report by Dr. Depaul ; the second reference is to an independent report communicated to the Prefect of Vannes by Dr. Bodelio of L'Orient, who knew the circumstances among the Breton peasantry at first hand.

vaccinating midwife, and on that day * furnished lymph for 104 direct arm-to-arm vaccinations in the following order: 17 at Brandivy, 26 at Camors, 31 at Plumergat, and 30 at Sainte Anne Plumeret. On the 12th of June the same midwife took lymph from two of the latter series, and vaccinated therewith 23 others at the village of Plumeret. Before following the fate of these 127, it will be necessary to go back to the original vaccinifers.

The child Mahé, aged five months, vaccinated on 21st May with the Vannes lymph *sur plaque*, became very ill in due course; the vesicles broke and continued as open sores for seven weeks. When seen on 19th August the infant was found to be quite well, with two ordinary scars on one arm and three on the other, and with a slight indolent enlargement of the glands in the armpits. The child Norcy, aged ten months, vaccinated the same day, was also ill in consequence of vaccination, the vesicles breaking and continuing as sores for five weeks; three weeks after vaccination he had a general reddish eruption. When seen on 19th August by the commissioners from Paris, he was found to be a large fat infant at the breast, with two scars on each arm, still rather red, a small amount of indolent axillary swelling, but with his skin perfectly free from eruption. He had been the vaccinifer of Marie Rosnaro, whose twelve vesicles furnished lymph for the series of 104 on the 5th June. On the 20th August the infant Rosnaro was found by the examining commissioner to be perfectly well, with six marks on each arm, correct as regards size and colour, and with no trace of axillary swelling or skin eruption; her vaccination was reported to have been regular from first to last; her parents were free from illness, except that the mother was in bed with rheumatism.

More than half (it is doubtful what the exact proportion was) of the 104 children vaccinated in succession from Rosnaro's vesicles at the villages of Brandivy, Camors, Plumergat, and St. Anne Plumeret, had serious after-effects. When information of this misadventure

* The statement originally made to the Academy that lymph was taken from this child on three successive days (3rd, 4th, and 5th of June) was not confirmed; all the detailed cases, at least, had been vaccinated on the 5th, and that is the only date mentioned by Bodelio.

came in from all sides, Dr. Bodelio of L'Orient went over the ground exactly in the footsteps of the midwife, and learned the following particulars: The post-vaccinal evils were worse in those who had been vaccinated at the end of the day's round than in those who were reached early in the progress; the vesicles began to ulcerate from the 12th to the 15th day; they were converted into large, deep, crater-like sores on a hard base, sometimes distinct, sometimes two or three confluent and measuring up to two inches in length; the skin around them was congested; sores not covered by crusts had the character of phagedena; roseola on the thighs, buttocks, and other parts occurred in a large number; pemphigus of the hands and feet in three, mucous *plaques* of the lips in four, and, in two of the latter, fissures at the angles of the mouth; two or three of the adults who suckled the infants had sores of the nipples. The scars on the infants' arms when seen by Dr. Bodelio were half an inch to three-quarters of an inch in diameter, where not confluent, and they suggested in various ways the sequel of a "chancre rongeur."

Forty-two of the one hundred and four vaccinated on 5th June were visited by MM. Depaul and Henri Roger from Paris on the 19th and 20th August; at that date they found no sores open, but in many cases traces of induration of the scars, indolent swelling of the axillary and cervical glands, roseola or papules of the skin, in one case ulcerated tonsils, in another the remains of an enormous abscess in front of the chest, and in still another the traces of a large abscess in the armpit; in no cases condylomata *circa anum;* in several of the forty-two visited, there were no traces of post-vaccinal effects of whatever kind, nor any history of them; anti-syphilitic treatment had been very generally resorted to.

The second series of 24 cases, vaccinated on 12th June from two infants in the series of 104 (5th June), had a still larger proportion of casualties; 17 were visited, among which it may be noted that Nos. 8 and 9 had mucous *plaques* on the tonsils.

For these disasters M. Depaul blamed the lymph sent on May 20th from the Prefecture of Vannes, having no reason to suspect it of syphilis, but being evidently at a loss to find

another source for the train of misfortunes. M. Bodelio
was equally perplexed, but he was inclined to say rather :
" ce serait à faire doubter de la nature syphilitique de cette
déplorable vaccination." The points to keep in mind are
that the two infants Mahé and Norcy, vaccinated directly
with the Vannes lymph, had merely ulcerated vesicles, which
would have passed without comment but for the inquiry
raised on other grounds, and had no syphilitic after-effects ;
while the infant Rosnaro, who furnished lymph for the
series of 104, had neither sores nor constitutional symptoms,
but remained in all respects perfectly well. On the other
hand, her six vesicles on each arm were used on the ninth
day ; and the evil results of the vaccine furnished by her
were plainly seen to be greater the more the lymph ran dry,
just as in the Coblenz case, and in the instance of Mr.
Hutchinson's first series. Lastly, the constitutional effects
were mild in most of the cases ; it is only in the second
series that we hear of mucous *plaques* of the tonsils, in two
cases about three months from the date of vaccination.

The rational theory of the Morbihan disaster is that
ulceration, followed by induration and (or) phagedena, is
part of the natural history of cow-pox infection ; that it
is nearly always latent, or kept in check ; that in some
circumstances it may be brought out or reverted to ; that
those circumstances in the particular epidemic were the
date and number of vesicles raised on the vaccinifer, and
the draining of their lymph to the last drop, so as to
vaccinate an enormous number ; and, lastly, that a continuous reproduction of lymph from that stock tended to
confirm and even to intensify the reawakened powers
of the cow-pox matter, as evidenced by the more decided

"syphilitic" character of the secondaries (mucous patches on the tonsils) in two cases of the last group.

One other outbreak recorded in the "Bulletin" of the Paris Academy * may be briefly referred to.

In the Department of the Lot, in August, 1866, M. Lafaye of Cardeillac received vaccine lymph which had been taken by a neighbouring public vaccinator from the arm of a robust infant; with that he proceeded to vaccinate a healthy infant named Mas, aged three months, who developed correct vesicles and had no ill consequences. On two successive days, the 19th and 20th of August, he vaccinated direct from the arm of Mas twenty-two other children. The vesicles in thirteen of these did badly, the open sores remaining on the arms for two months, with constitutional symptoms. MM. Clary and Guary, who were sent some months after as a commission to report, found the infant Mas healthy; they confirmed the facts about the thirteen seriously infected from him, and detected condylomata *circa anum* in most of them.

In this series it is again to be noted that a large number were vaccinated from the same arm on two successive days; but unfortunately we have not the means of judging whether the nine who escaped the ill-effects were all or most of them vaccinated on the first day, and the thirteen who felt the ill-effects vaccinated all or most of them on the second day; also we are uninformed on the most essential point of the date after vaccination, or the period of maturity, at which the lymph began to be taken from Mas.

Down to 1869 the subject of post-vaccinal accidents continued to be debated in the Paris Academy of Medicine, and various other cases, for the most part of isolated

* *Loc. cit.*, sitting of 28th of February, 1867.

occurrence, and therefore more open to hypothetical allegations, were reported to it; but after that date the subject seems to have dropped, and we find no further reference to it until 1884, when a case is reported from the country at the sitting of 9th September, and referred to the Vaccination Committee, without the details being given subsequently in the " Bulletin." *

Of recent vaccinal disasters on the large scale, one of the most notable is the outbreak on the 30th December, 1880, in Algiers, wherein fifty-eight recruits of the 4th regiment of Zouaves were infected with the same disease as in the foregoing cases, and under corresponding circumstances.† Another is the painful Swiss outbreak of 1878, in which the victims were school girls.‡ But, on the whole, the literature of medicine during the last ten years has not added much to the record of vaccinal syphilis in groups of cases. In all probability such epidemics as those of Rivalta and Morbihan occur but rarely.

On the other hand, it is clear that isolated instances of

* I shall merely give references to other cases in the *Bulletin de l'Académie de Médecine :* Hérard, xxviii., 1862-63, p. 1,189 ; Millard, xxxii., 1866-67, p. 1,048 ; Schuh, *ib.,* p. 1,058 ; Alph. Guérin, xxxiv., 1869, p. 512 (one out of forty vaccinated at the same time : one of three vesicles became a crater-like indurated ulcer, followed by mucous plaques *apud vulvam*); Chassaignac (case of 1863), *ib.* p. 783 ; Zallonis, *ib.* p. 1,017 ; Vicherat, *ib.* 1,103 ; Bardinet, *ib.* 1,171.   Depaul's "Projet de Rapport," in vol. xxx. (1864-65), gives a summary of the earlier and foreign cases up to date.

† *Journal de Hygiene,* 25th of August, 1881.   The lymph for re-vaccinating the recruits was taken from two infants not quite two months old, who looked perfectly healthy.

‡ *Bulletin de la Société Medicale de la Suisse romande,* quoted by Fournier, *op. cit.,* 1886, p. 590.

vaccinal ulcer, in no respect differing as regards primary characters from the sore arms in an epidemic series, are somewhat commonplace incidents. In the second series reported by Mr. Hutchinson, there were eleven children suffering from after-effects of vaccination, that were reckoned equivalent to syphilis, and yet nothing was thought of it in each individual case until two of their number attended at a hospital in the East End of London several months after. It would be quite misleading to estimate the number of such cases by the infrequency of the epidemic outbreaks; a series involving many families in one neighbourhood at the same time will, of course, make most stir, and secure an official inquiry; but there must be many more single cases where the public vaccinator loses all trace, and in which the sequence of events is observed by no other medical man until after an interval long enough to make the link of cause and effect uncertain or improbable. Do these cases present themselves afterwards as cases of infantine syphilis? and, if so, have we any means of tracing them on the large scale to vaccination?

This brings me to consider the very extraordinary and disproportionate increase in the death-rate from syphilis among infants in their first year in England and Wales; and as that striking fact (one of the most striking, in my opinion, in the whole of the Registrar-General's tables during the last thirty years) has nowhere been fully faced and grappled with, so far as I can discover, by the Registrar-General himself, or by the medical press, while, on the other hand, it has been adduced in Parliament and elsewhere as convincing evidence of the dangers of vaccination, I shall give a separate section to a dispassionate inquiry

whether the increase, by leaps and bounds, of the infantine death-rate from syphilis during the last forty years
can be explained in whole or in part without resorting
to the unwelcome hypothesis of the infective cow-pox
sore.

CHAPTER IX.

THE INCREASING DEATH-RATE FROM INFANTINE SYPHILIS.

THE registration returns are not published continuously
farther back than the year 1847, and the abstracts of the
causes of mortality at different periods of life do not
regularly begin until 1855. The table of syphilis mortality among infants in their first year on the one hand,
and at all other ages on the other, cannot therefore be
given so as to cover completely the period of compulsory
vaccination, which begins with 1854. But there are
various ways of arriving at an approximate estimate of
the incidence of the mortality previous to 1855 in the
respective periods of life : one of these is the table,
standing by itself, for the year 1847 ; another is Dr.
Farr's sample table of the causes of death at various ages,
for the female sex alone, in 1852 ; and a third is the
continuous table from 1847 (as well as of two previous
years) for London alone. I begin with a series of years
of the last-mentioned, in order to show that there is no
improbability in the estimate of the infantine death-rate
for all England in the years immediately preceding
1855.

Deaths from Syphilis in London, 1843—1858.

	Infants under one year.	All other ages.		Infants under one year.	All other ages.
1843	24	45	1851	95	31
1844	25	56	1852	101	30
1845	—	—	1853	112	40
1846	—	—	1854	133	73
1847	66	62	1855	134	45
1848	72	50	1856	160	56
1849	62	41	1857	181	46
1850	79	46	1858	196	70

Increasing Infantine Death-rate from Syphilis. (England and Wales.)

	Infants under one year.	All other ages.		Infants under one year.	All other ages.
1847	255	310	1866	1,180	482
1848	—	—	1867	1,241	457
1849	—	—	1868	1,364	522
1850	—	—	1869	1,361	498
1851	—	—	1870	1,422	436
1852	380	243	1871	1,317	425
1853	380	242	1872	1,410	421
1854	591	373	1873	1,376	467
1855	579	368	1874	1,484	513
1856	579	300	1875	1,554	580
1857	656	301	1876	1,580	554
1858	684	322	1877	1,550	524
1859	778	311	1878	1,647	535
1860	767	300	1879	1,493	536
1861	798	379	1880	1,588	571
1862	867	378	1881	1,540	557
1863	983	403	1882	1,666	561
1864	1,089	461	1883	1,813	500
1865	1,155	492	1884	1,733	547

The tables show an enormous and steady increase in the deaths of infants; while in the deaths at all ages above one

J

year, the rise is hardly proportionate to the increase in
the numbers living at those ages.   The increase is greatest
in 1854, which happens to have been the first year of compulsory vaccination ; and, in so far as the ages above one
share the sudden leap, an inspection of the London tables
brings out the curious fact that it is mainly due to an unusual number of deaths about the years of puberty.

The first remark to be made upon this statistical result
is that there may be a source of fallacy in it—that the increase may be only apparent.   Better diagnosis, it is said,
has led to the causes of infantine mortality being better
discriminated.   Greater conscientiousness, it is also said,
has led to a more correct return on the part of the certify-
ing practitioner.*   Furthermore it is contended that the
death-rate from syphilis in the later periods of life does not
serve to measure the full prevalence of that disease, that
thousands of persons with constitutional lues die of some
intercurrent disease and have their death registered under
the head of the latter, whereas in infancy the syphilitic
who die are mostly ranged under syphilis.   I shall now
consider these arguments.

That the death-rate from a disease in the Registration
returns should be often a fallacious measure of the prevalence of that disease is undisputed ; that an inherited taint
may kill the child while the progenitor is spared, is also a
reasonable contention.   At the same time it has to be kept
in mind that a large number of infants born with that taint
will be cut off, in the ordinary course of events, by some
one of the various maladies so destructive of infant life, such
as diarrhœa, whooping-cough, or those nondescript forms of

* *Report of the Registrar-General for* 1879, Lond., 1881.

teething troubles which have thousands of deaths put down to them every year in the mortality returns. If intercurrent disease claims the syphilitic in the other periods of life, it will claim a proportion of them in infancy, and probably as large a proportion of them.

Again, the Registrar-General has associated with better diagnosis the other factor of greater conscientiousness in making the returns. But how does that improvement affect the statistics of infantine disease more than those of adult maladies? The Registrar-General himself shall answer. In the first place, the Reports for several years dwell upon the fact that, in infantine syphilis, the sins of the parents are visited upon the children. In the next place, the Registrar-General, in his Report for 1883 (p. xvi.), commenting on the negligence of medical men, especially in the provinces, to state on their certificates, in the case of death from small-pox, whether the child or other person had been vaccinated or not,* expresses regret that " medical men in the country should not take pains to ascertain this fact, and state it fearlessly in their certificates. . . . . The private

---

* To show the grounds of the Registrar-General's complaint, I have drawn up the following table from the Returns for four years :—

Deaths from Small-Pox in Infants under One Year, in the Provincial Districts of England and Wales.

Year.	Total Deaths under one.	Unvaccinated.	Vaccinated.	Not stated.
1881	66	13	1	52
1882	90	20	1	69
1883	108	21	1	86
1884	146	56	5	85

medical attendant is apparently unwilling to state facts
which might be disagreeable to the relatives of the de-
ceased." If that be really a motive, it is a motive that
must operate most strongly of all where the practitioner
believes that the malady is a reminiscence of the sins of
the parent. In so far as the Registration Office has any
theory of the increase of infantine syphilis, the reasoning is
not always consistent with official doctrine as applied in
other instances. There are other grounds for not regarding
the increase shown in the infantine column of the table
as merely due to more accurate or more conscientious
returns : the chief of these, and a reason quite sufficient in
itself, being that the increase, after the great leap in 1851,
has been steadily progressive, whereas in the column for
all other ages the figures have undergone little change
during the last twenty years, notwithstanding the large
increase of urban population.

It will be necessary to assume that the steady increase
shown by the returns in the deaths from syphilis of infants
in their first year is a real increase. The question then
arises, whether transmission from one or both parents can
account for the whole of it. The subject of infantine
syphilis is beset with exceptional difficulties, as any one conversant with medical writings must be aware. It has
occasioned some of the most paradoxical positions in recent
medical literature, such as the contention of the late Professor Parrot that hereditary syphilis is the sole cause of so
common a disease as rickets,* and the argument of a writer

---

* "Le rachitis reconnait pour cause unique la syphilis héréditaire."
Trans. Internat. Med. Congress, 1881. Lond., 1881, iv. 35. See also the
posthumous volume, *La syphilis héréditaire et le rachitis.* Paris, 1886.

in a Vienna journal,* that there is no such thing as hereditary syphilis at all. Paradoxes of that kind arise out of the confusion in which infantine syphilis is involved. In a former treatise,† I indicated a way out of this confusion. The error of growth in the bones, which is the central fact of rickets, was traced deductively to defective structure and function of the great extemporised organ of intra-uterine nutrition, the placenta ; and as we know that the same organ suffers (sometimes very obviously) in constitutional lues of the mother, those elements in the congenital lues of the child, which have an undoubted parallelism with rickets, were ascribed to the same placental failure, and were accordingly reckoned as an intelligible result of deficient fœtal endowment or intra-uterine nutrition, due to a special cause but not due to infection at all. The acceptance of that hypothesis would at once get rid of some of the chief perplexities in infantine syphilis, and, among the rest, of all that is paradoxical in M. Parrot's position.

Moreover, it would lead to some sort of classification or subdivision of the generic total of syphilis in infants and children, including the inherited (in the stricter sense), the congenital, and that due to contagion. It would be hazardous, however, to attempt a strict line of demarcation between the proper effects of intra-uterine or placental malnutrition, and the effects due to heredity in the ordinary sense. The best guide to follow would be the familiar fact

---

* Hermann, " Die Vererbung der Syphilis," *Wiener Allgem. Med. Ztg.*, 1882. No. 19—23.

† Article "Pathology," *Encycl. Britan.* xviii., 1885: section on " Placental Function in Congenital Disorders," p. 374, and on "Syphilis of the Offspring," p. 405.

that heredity in all other constitutional diseases, even if they have an infective character as well, is not apt to be manifested until infancy is past. Hereditary syphilis in that sense, as distinguished from the congenital, has found a more than theoretical position during recent years in the numerous cases described as late inherited syphilis, or the " syphilis héréditaire tardive," of Fournier ; * whereby are meant those cases in which the phenomena come out after the first infancy, during adolescence, and even in adult life, or, in fact, at the periods when other inherited constitutional disorders are apt to come out. Contrasted with the "syphilis héréditaire tardive " is the "syphilis héréditaire précoce," which is seen immediately after birth, or very soon after, and may or may not have existed in those who afterwards manifest the tardy signs of the disease. Supposing, however, that heredity, in the strict sense, were the right factor to blame for deaths from syphilis under the age of one, there does not appear to be any known law of morbid inheritance which would account for a taint in the progeny increasing out of all proportion to its prevalence in the parents—to the extent, indeed, of four-fold in the course of little more than thirty years.

On the other hand, if feeble placental power on the part of the mother be recognised as responsible for that part of infantine syphilis which is closely parallel with rickets (including the condition of the liver and spleen as well as the errors of growth in the bones and teeth), it will not be difficult to find a cause of increase in recent times, apart from vaccinal syphilis; which latter would, at all events, be insufficient to account for all those cases of infantine syphilis

* *La syphilis héréditaire tardive.*   Paris, 1886.

where the bone-lesions are tolerably well marked. Whatever things in the health of the mother tend ordinarily to future rickets in the child that she is pregnant with (they are numerous and not absolutely confined to the poor), the same would have the effect of making more certain that feebleness or failure of placental function and structure which is a not uncommon incident of constitutional syphilis of the mother (as measured by the frequency of miscarriages), even amidst favourable circumstances of living. In that way, and still holding to the special or specific cause of placental mal-nutrition, we might connect the recent increase of congenital syphilis, supposing it distinguishable from inherited, with something operating on a considerable scale among child-bearing women, and taxing their maternal functions in a general way. It is easy, of course, to go wrong in assigning such general causes; but it can hardly be doubted that the enormous development of industrial competition, affecting the female sex as well as the male, is a factor of the kind here sought for. A relevant physiological subject, which I have given a good deal of attention to,* is the periodical building up, by the tissues of the mother, of the great organ of intra-uterine nutrition, the placenta. Whoever goes minutely into the details of that extemporised formation, can hardly fail to be impressed by the vast constructive power that must be implicitly present whenever the occasion arises, so as to secure efficient nutrition for the child; and one may well marvel that the building up of the placenta is done on the whole so well, and that the supplies of that glandular organ are provided so adequately for the offspring in the womb,

* *Journal of Anatomy and Physiology*, July, 1878, and January, 1879.

considering the circumstances of many child-bearing women in the industrial class and among the poor.

If the formative placental powers of the mother are already impaired by the specific taint in her system, according to what we otherwise know of constitutional syphilis; then it is easy to understand that an aggravation of the general circumstances that lead to placental mal-nutrition, and, as an ordinary result, to rickets in the offspring, would reveal its effects among the offspring of syphilitic mothers by an increase in the cases of infantine syphilis. The effects in the child would not be hereditary syphilis, but congenital syphilis; just as rickets is a congenital, but by no means a hereditary malady in the strict meaning of heredity.

There would still remain the third of the factors or elements which I have assumed as entering into the composite disease called infantine syphilis, namely, direct infection of the offspring by the mother. In what way infection in the course of pregnancy or during parturition can produce secondary symptoms in the child (without the usual primary sore), namely the rashes, excoriations, mucous thickenings and tubercles, and the like; or how such infection may co-mingle its effects with those of congenital mal-nutrition (itself due to a specific cause on the mother's side), so as to produce the most complete type of infantine syphilis—these are questions which are confessedly obscure.* In so far as the intra-uterine infective element can

---

* To show that the syphilis of the new-born is recognised as a difficult problem, and to exemplify the kind of solutions suggested for it, I may refer to the short paper on "The Origin of Infantile Syphilis," by Dr. R. Cory, *Lancet*, 1876, i. 885.

be considered apart from that of congenital mal-nutrition, it is not easy to see why it should have undergone so great and steady an increase during the last thirty or forty years as the tables of mortality for the first year of life point to. It is probably the case, however, that a great part of the increase of the deaths in the first year will have to be looked for in the working of those two factors, before we go to vaccination as a last resort.

This is made all the more likely by an examination of the more recent registration returns for Scotland, which give the mortality in separate columns for the first three months of life, the second three months, and the next six months; if these returns may be taken as showing forth the incidence upon the several quarters of the first year for England also, we should have the significant fact that considerably more than half the deaths from syphilis under the age of one belong to the first three months of infancy. That fact certainly does not make absolutely against the hypothesis of an increase being due to vaccinal infection ; for we find that four of Whitehead's least ambiguous cases had their symptoms directly following vaccination at the age of one month or two months.

As a matter of fact, there are many authenticated cases, some of them fatal, belonging either to epidemics or occurring singly, where rashes, mucous tubercles, marasmus, and the like have followed primary vaccinal sores, through no complication of venereal syphilis, either actually proved or hypothetically intelligible, but simply because the cow-pox, in respect to its original although mostly latent characters, runs on all fours with the venereal pox itself. In the analysis, then, of the composite total of infantine syphilis,

I should ascribe part of the increase to the congenital or placental factor, as distinguished from the strictly hereditary, and to the obscure element of direct infection or contagion more or less constantly co-mingled therewith; but, on the evidence of facts, I should ascribe some part also to the infection of cow-pox in and by itself. It will now remain to state briefly the argument from the registration returns, which makes that claim on behalf of cow-pox an admissible one.

A sore on the arm after vaccination is not uncommon; it will probably be found to be rather commoner than "erysipelas after vaccination," from which cause the deaths among infants under one year in England and Wales from 1855 to 1880 were as in the following table (the deaths from simple "erysipelas" are given in the adjoining column for comparison) :

Deaths from Erysipelas after Vaccination, and from Erysipelas, under one year of age.

	Erysipelas after Vaccination.	Erysipelas.		Erysipelas after Vaccination.	Erysipelas.
1855	0	583	1870	20	685
1856	5	610	1871	22	716
1857	0	421	1872	16	617
1858	0	599	1873	19	675
1859	5	569	1874	27	867
1860	2	514	1875	36	796
1861	2	492	1876	21	700
1862	3	458	1877	26	667
1863	7	612	1878	35	582
1864	11	618	1879	31	561
1865	10	579	1880	32	618
1866	9	527	1881	56*	644
1867	3	467	1882	65*	696
1868	8	647	1883	51*	641
1869	19	589	1884	49*	618

* "Cow-pox and other effects of vaccination."

In the Registration tables the form of entry was changed, after 1880, from "erysipelas after vaccination" to "cow-pox and other effects of vaccination;" at the same time the cases nearly doubled under the new heading. About one-half of these recent deaths, then, we may ascribe to "cow-pox," as distinguished from the erysipelas incidental to it, although the term probably stands for a rather composite group in the certificates returned to Somerset House. The entry is a new one; if it should grow (in the returns) as remarkably as "erysipelas after vaccination" has grown, from zero to some thirty cases annually, we may expect that cow-pox will one day be publicly charged with a considerable total of deaths. That it would be charged with more deaths and sickness than at present, if its inherent nature and latent possibilities were better known, is, I think, highly probable. So long as cow-pox is supposed to be small-pox of the cow, there can be no vigilant outlook for or correct appreciation of its consequences in the weeks, or months, or years following.

The real affinity of cow-pox is not to the small-pox but to the great pox. The vaccinal roseola is not only very like the syphilitic roseola, but it means the same sort of thing. The vaccinal ulcer of every-day practice is, to all intents and purposes, a chancre; it is apt to be an indurated sore when excavated under the scab; when the scab does not adhere, it often shows an unmistakable tendency to phagedena. There are doubtless many cases of it where constitutional symptoms are either in abeyance, or too slight to attract notice. But in other instances, to judge from the groups of cases to which inquiry has been mostly directed, the degeneration of the vesicle to an indurated or phagedenic sore (all in its day's work) has been followed by roseola, or by scaly and even pemphigoid eruptions, by iritis, by raised patches or sores on the tonsils and other parts of the mouth or throat, and by condylomata (mucous tubercles) elsewhere.*

Those who believe that such after-effects are the exclusive prerogative of venereal pox will, of course, vehemently contest this view of the matter. The appeal must be in the end to facts ; and a careful and unbiassed survey of the facts has convinced me that cow-pox sores must be credited with a power of producing secondary symptoms (I say nothing of tertiary), not because they have the contamination of venereal pox in them, but because their nature is the same as or parallel with that of the venereal pox itself. The unmentionable circumstances of the latter are not the only occasion of sores acquiring inveteracy and a long train of effects perpetuated and intensified by reproduction through a succession of cases. The natural history of cow-pox, which I have said enough of in earlier chapters, tells the same story under circumstances totally different.

The rational view of cow-pox appears to me to be made much easier for the intelligence and belief, by discovering a corresponding rational origin for the specific characters of the venereal pox itself. But the rational view of cow-pox need not stand or fall with the other piece of rationalism.

---

* Affections of the bones have not been noticed in any of the epidemics of vaccinal syphilis. In the Styrian epidemic there were somewhat vague indications of infection of the lungs and kidneys in one or two cases. If vaccinal syphilis were reproduced through as many generations as the venereal pox has been, it would probably breed tertiary effects in the same small proportion of cases as the latter. But it is only in rare instances that the ulcerative effects have been reproduced through two or more removes.

Apologists for the *ab æterno* specificity of venereal pox may find some plausible ground for declining to entertain a rationalistic or common-sense explanation of the inveteracy, reproductiveness, and constitutional infectiveness of that disease. But no such obscurantism is possible in the case of cow-pox. The whole facts and circumstances are clearly before anyone who has the eyes to see. The first duty of everyone is once for all to disabuse his mind of Jenner's invention of the name *variolæ vaccinæ* for cow-pox. The affection of the cow's udder was long recognised by common folks as a pox in the original and classical English sense of the word; the name of it in Norfolk was pap-pox. No one had dreamt of discovering any resemblance in it to the pustules of the foreign contagious skin-disease which came to be called the small-pox, until Jenner, by a master-stroke of boldness and cunning, placed the Latin name *variolæ vaccinæ* first on his title-page,* as if he were merely expressing in scientific form the universally accepted meaning of the colloquial name. There was no candid or overt attempt, in the body of his essay, to justify that daring innovation ; most of his readers from that time to this have hardly realised that it was an innovation at all, for the reason that Jenner adroitly left his title-page to justify itself. His trumped-up name somehow passed without challenge, except for a grammatical objection on the part of Pearson, and a general criticism by Moseley ; and although the want of likeness, still more in circumstances than in form, between the pustules of small-pox and even the modified kind of inoculated cow-pox vesicle has been

---

* *Inquiry into the Causes and Effects of Variolæ Vaccinæ, a Disease known by the name of the Cow-Pox.* Lond., 1798.

pointed out in elaborate detail by several writers, and ought, indeed, to be so obvious to anyone as not to need pointing out at all; yet the Jennerian fable of *variolæ vaccinæ* continues to be the creed of the medical profession.

The first thing, then, is to dismiss Jenner's Latin name for cow-pox. Having dismissed the Latin name we come to the thing itself—to its characters on the milch-cow's teats, and the circumstances of its very occasional origin thereon; to its characters as communicated by contact to the milker's hands and face; to its characters as experimentally cultivated and modified by artificial selection on the arms of infants; and to the occasional reversions of type, with disastrous consequences, in the ordinary course of vaccination practice. These are the things that I have endeavoured to set forth in the several chapters of this essay. I appeal to facts that are as well authenticated as any facts can be, and I invite the most rigid scrutiny of my use of them, or of my reasoning from them. I deprecate no criticism; but I warn the apologists of the Jennerian doctrine that any attempt to wrap themselves in a mantle of orthodoxy will be a grave dereliction of that duty which the profession owes to the public. I am as sensible as any one of the need of securing our professional credit and dignity in the controversy which has been raised, by an intelligent and ceaselessly active body of the laity, touching the whole subject-matter of compulsory vaccination; and it is because I am persuaded that the profession must lead rather than follow public opinion in bringing the theory of cow-pox up to date, that I have thrown this contribution to the subject into a strictly professional and even technical form.

INDEX.

PRINTED BY CASSELL & COMPANY, LIMITED, LA BELLE SAUVAGE, LONDON, E.C.